INTRODUCTION TO QUANTUM OPTICS
From Light Quanta to Quantum Teleportation

The purpose of this book is to provide a physical understanding of what photons are and of their properties and applications. Special emphasis is made in the text on photon pairs produced in spontaneous parametric down-conversion, which exhibit intrinsically quantum mechanical correlations known as entanglement, and which extend over manifestly macroscopic distances. Such photon pairs are well suited to the physical realization of Einstein–Podolsky–Rosen-type experiments, and also make possible such exciting techniques as quantum cryptography and teleportation. In addition, non-classical properties of light, such as photon antibunching and squeezing, as well as quantum phase measurement and optical tomography, are discussed. The author describes relevant experiments and elucidates the physical ideas behind them. This book will be of interest to undergraduates and graduate students studying optics, and to any physicist with an interest in the mysteries of the photon and exciting modern work in quantum cryptography and teleportation.

HARRY PAUL obtained a Ph.D. in Physics at Friedrich Schiller University, Jena, in 1958. Until 1991 he was a scientific coworker at the Academy of Sciences at Berlin. Afterwards he headed the newly created research group Nonclassical Light at the Max Planck Society. In 1993 he was appointed Professor of Theoretical Physics at Humboldt University, Berlin. He retired in 1996.

Harry Paul has made important theoretical contributions to quantum optics. In particular, he extended the conventional interference theory based on the concept of any photon interfering only with itself to show also that different, independently produced photons can be made to interfere in special circumstances. He was also the first to propose a feasible measuring scheme for the quantum phase of a (monochromatic) radiation field. It relies on amplification with the help of a quantum amplifier and led him to introduce a realistic phase operator.

Harry Paul is the author of textbooks on laser theory and non-linear optics, and he is editor of the encyclopedia *Lexikon der Optik*.

INTRODUCTION TO QUANTUM OPTICS

From Light Quanta to Quantum Teleportation

HARRY PAUL

Translated from German by
IGOR JEX

CAMBRIDGE
UNIVERSITY PRESS

CAMBRIDGE
UNIVERSITY PRESS

University Printing House, Cambridge CB2 8BS, United Kingdom

Cambridge University Press is part of the University of Cambridge.

It furthers the University's mission by disseminating knowledge in the pursuit of
education, learning and research at the highest international levels of excellence.

www.cambridge.org
Information on this title: www.cambridge.org/9780521835633

German edition © B. G. Teubner GmbH, Stuttgart/Leipzig/Wiesbaden 1999

English translation © Cambridge University Press 2004

First published in the German language as *Photonen. Eine Einführung in die Quantenoptik,
2. Auflage* [2nd Edition] by B. G. Teubner GmbH 1999. English translation published 2004

A catalogue record for this publication is available from the British Library

Library of Congress Cataloguing in Publication data

Paul, Harry.
[Photonen, English]
Introduction to quantum optics: from light quanta to quantum teleportation / Harry Paul;
translated from German by Igor Jex.
p.   cm.
Includes bibliographical references and index.
ISBN 0 521 83563 1
1.   Photons.   I. Title.
QC793.5. P42P3813 2004
539.7'217 – dc22   2003062729

ISBN 978-0-521-83563-3 Hardback

To Herbert Walther, with heartfelt gratitude

Contents

Preface

All the 50 years of conscious pondering did not bring me nearer to the answer to the question "What are light quanta". Nowadays every rascal believes, he knows it, however, he is mistaken.

Albert Einstein
(1951 in a letter to M. Besso)

The rapid technological development initiated by the invention of the laser, on the one hand, and the perfection attained in the fabrication of photodetectors, on the other hand, gave birth to a new physical discipline known as quantum optics. A variety of exciting experiments suggested by ingenious quantum theorists were performed that showed specific quantum features of light. What we can learn from those experiments about the miraculous constituents of light, the photons, is a central question in this book. Remarkably, the famous paradox of Einstein, Podolsky and Rosen became a subject of actual experiments too. Here photon pairs produced in entangled states are the actors.

The book gives an account of important achievements in quantum optics. My primary goal was to contribute to a physical understanding of the observed phenomena that often defy the intuition we acquired from our experience with classical physics. So, unlike conventional textbooks, the book contains much more explaining text than formulas. (Elements of the mathematical description can be found in the Appendix.) The translation gave me a welcome opportunity to update the book. In particular, chapters on the Franson experiment and on quantum teleportation have been included.

I expect the reader to have some knowledge of classical electrodynamics, especially classical optics, and to be familiar with the basic concepts of quantum theory.

I am very grateful to my colleague Igor Jex from the Technical University of Prague, who was not discouraged from translating my sometimes rather intricate German text. (Interested readers may like to consult Mark Twain's "The Awful German Language" in *Your Personal Mark Twain* (Berlin, Seven Seas Publishers, 1960).)

Harry Paul
(Berlin, September 2003)

1

Introduction

And the Lord saw that the light was good.
Genesis 1:4

Most probably all people, even though they belong to different cultures, would agree on the extraordinary role that light – the gift of the Sun-god – plays in nature and in their own existence. Optical impressions mediated by light enable us to form our views of the surrounding world and to adapt to it. The warming power of the sun's rays is a phenomenon experienced in ancient times and still appreciated today. We now know that the sun's radiation is the energy source for the life cycles on Earth. Indeed, it is photosynthesis in plants, a complicated chemical reaction mediated by chlorophyll, that forms the basis for organic life. In photosynthesis carbon dioxide and water are transformed into carbohydrates and oxygen with the help of light. Our main energy resources, coal, oil and gas, are basically nothing other than stored solar energy.

Finally, we should not forget how strongly seeing things influences our concepts of and the ways in which we pursue science. We can only speculate whether the current state of science could have been achieved without sight, without our ability to comprehend complicated equations, or to recognize structures at one glance and illustrate them graphically, and record them in written form.

The most amazing properties, some of which are completely alien to our common experiences with solid bodies, can be ascribed to light: it is weightless; it is able to traverse enormous distances of space with incredible speed (Descartes thought that light spreads out instantaneously); without being visible itself, it creates, in our minds, via our eyes, a world of colors and forms, thus "reflecting" the outside world. Due to these facts it comes as no surprise that optical effects confronted our knowledge-seeking mind with more difficult problems than those of moving material objects. Over several hundred years a bitter war was fought between two parties. One group, relying on Newton's authority, postulated the existence of elementary constituents of light. The other, inspired by the ideas of Huygens, fought for light as a wave phenomenon. It seemed that the question was ultimately settled in favor of the wave alternative by Maxwell's theory, which conceived light as a special form of the electromagnetic phenomena. All optical phenomena could be related without great difficulty and to a high degree of accuracy to special solutions of the basic equations of classical electrodynamics, the Maxwell equations.

However, not more than 40 years passed and light phenomena revealed another surprise. The first originated in studies of black-body radiation (radiation emitted from a cavity with walls held at a constant temperature). The measured spectral properties of this radiation could not be theoretically understood. The discrepancy led Max Planck to a theory which brought about a painful break with classical physics. Planck solved the problem by introducing as an *ad hoc* hypothesis the quantization of energy of oscillators interacting with the radiation field.

On the other hand, special features of the photoelectric effect (or photoeffect) led Einstein to the insight that they are most easily explained by the "light quantum hypothesis". Based on an ingenious thermodynamic argument Einstein created the concept of a light field formed from energy quanta $h\nu$ localized in space (h is Planck's constant and ν is the light frequency).

The newly created model was fully confirmed in all its quantitative predictions by studies of the photoeffect that followed, but there was also no doubt that many optical phenomena like interference and diffraction can be explained only as wave phenomena. The old question, Is light formed from particles or waves?, was revived on a new, higher level. Even though painful for many physicists, the question could not be resolved one way or the other. Scientists had to accept the idea that light quanta, or photons as they were later called, are objects more complicated than a particle or a wave. The photon resembles a Janus head: depending on the experimental conditions it behaves either like a particle or as a wave. We will face this particle–wave dualism several times in the following chapters when we analyze different experiments in our quest to elucidate the essence of the photon. Before this, let us take a short stroll through the history of optics.

2

Historical milestones

2.1 Light waves à la Huygens

While the geometers derive their theorems from secure and unchallengeable principles, here the principles prove true through the deductions one draws from them.

Christian Huygens
(Traité de la Lumiére)

Christian Huygens (1629–1695) is rightfully considered to be the founder of the wave theory of light. The fundamental principle enabling us to understand the propagation of light bears his name. It has found its way into textbooks together with the descriptions of reflection and refraction which are based on it.

However, when we make the effort and read Huygens' *Treatise of Light* (Huygens, 1690) we find to our surprise that his wave concept differs considerably from ours. When we speak of a wave we mean a motion periodic in space and time: at each position the displacement (think about a water wave, for instance) realizes a harmonic oscillation with a certain frequency v, and an instantaneous picture of the whole wave shows a continuous sequence of hills and valleys. However, this periodicity property which seems to us to be a characteristic of a wave is completely absent in Huygens' wave concept. His waves do not have either a frequency or a wavelength! Huygens' concept of wave generation is that of a (point-like) source which is, at the same time, the wave center inducing, through "collisions" that "do not succeed one another at regular intervals," a "tremor" of the ether particles. The given reason for wave propagation is that ether particles thus excited "cannot but transfer this tremor to the particles in their surrounding" (Roditschew and Frankfurt, 1977, p. 31). Therefore, when Huygens speaks of a wave, he means an excitation of the ether caused by a *single* perturbation in the wave centrum, i.e. a single wavefront spreading with the velocity of light. The plots drawn by Huygens showing wavefronts in an equidistant sequence have to be

3

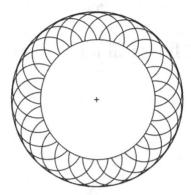

Fig. 2.1.   Propagation of a spherical wave according to the Huygens principle.

understood such that it is the *same* wavefront at different times, and the regularity of the plot is caused exclusively by having chosen identical time differences.

In fact, what was correctly described in this way is white light – sunlight for instance. The time dependence of the excitation – or more precisely speaking the electric field strength component with respect to an arbitrarily chosen direction – is not predictable but completely random (stochastic).

On the other hand, it is also clear that such a theory is not able to explain typical wave phenomena such as interference or diffraction where the wavelength plays an important role. It required Newton and his ingenious insight that natural light is composed of light with different colors to come nearer to an understanding of these effects.

This should not hinder us, however, from honoring Huygens' great "model idea" known as the Huygens principle according to which each point either in the ether or in a transparent medium reached by a wave, more precisely a wavefront, becomes itself the origin of a new elementary wave, as illustrated in Fig. 2.1 for the example of a spherical wave.

The wavefront at a later time is obtained as the envelope of all elementary waves emitted at the same, earlier, moment of time. However, Huygens could not answer the question of why a *backwards running* wave is generated only at the boundary of two different media and not also in a homogeneous medium (including the ether). In fact, a satisfactory answer could not be given until Augustin Fresnel complemented the principle with the principle of interference – we use today the term Huygens–Fresnel principle – the strength of which is demonstrated when we treat theoretically the problems of diffraction. By the way, the answer to the above question is simple: the backwards running waves "interfere away."

But let us return to Huygens! Using the assumption that light propagates in two different media with different velocities, we can easily explain reflection and

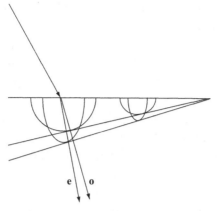

Fig. 2.2. Birefringence after Huygens. o = ordinary beam; e = extraordinary beam; the arrows indicate the beam direction.

refraction of light using the Huygens principle. The explanation of the strange effects of birefringence (the splitting of the incident beam into an ordinary and an extraordinary beam) can be viewed as an extraordinary success of the principle. It was based on the ingenious guess that the surfaces of propagation for the extraordinary beam are not spherical surfaces but the surfaces of a rotational ellipsoid (see Fig. 2.2), an insight that is fully verified by modern crystallography. However, Huygens had great difficulty understanding the experiments he performed with two crystals of calcite. We will discuss this problem in more detail in the next section.

2.2 Newton's light particles

> As in mathematics, so in natural philosophy, the investigation of difficult things by the method of analysis, ought ever to precede the method of composition. This analysis consists in making experiments and observations, and in drawing general conclusions from them by induction, and admitting of no objections against the conclusions, but such as are taken from experiments, or other certain truths.
>
> *Isaac Newton*
> (Opticks, *3rd Book*)

Isaac Newton (1643–1727) was the founder of the particle theory of light. Even though the light particles postulated by Newton do not have anything in common with what we now call photons, it is still exciting to trace back the considerations which led such a sharp mind to the conclusion that light of certain color is composed of identical, elementary particles. As an abiding supporter of the inductive method as the method of natural sciences, Newton was guided by a simple experience: the straight line propagation of light "rays," recognizable on the sharp contours of shadows of (nontransparent) objects placed into the beam path. This effect seemed to Newton to be easily explained by assuming that the light source emits tiny "bullets" propagating along straight lines until they interact with material objects. He believed a wave process to be incompatible with straight line propagation of the excitation. Water waves showed a completely different type of behavior: they obviously run around an obstacle!

Since the breakthroughs of Young and Fresnel it has been known that Newton's conclusion was premature. What happens when a wavefront hits an obstacle depends crucially on the ratio between the size of the obstacle and the wavelength. When the ratio is very large, the wave character is not noticeable; in the limit of very small wavelength the propagation of light can be described by straight lines, i.e. light rays. On the other hand, wave properties become predominant when the dimensions of the obstacle are of the order of the wavelength, as in the above example of water waves.

Newton himself observed with impressive precision experimental phenomena where the outer edge of a body (for instance the cutting edge of a razor) deflects the light "rays" in its proximity a little from their original straight direction so that no ideally sharp shadows are observable. He did not take these phenomena, now called the diffraction of light, to be hints of a wave-like character of light; instead, he considered the bending as the result of a force applied onto the particles (in his opinion caused by the density of the ether increasing with increasing distance from the object), an idea which was completely in accord with the well established concepts of mechanics. Newton's belief that light had a particle nature should be judged from the perspective of the seventeenth century atomism, which was, at the time, a deeply rooted concept. "True" physics – in contrast to scholasticism which categorized light and color phenomena into the class of "forms and qualities" – was imaginable only as a mechanical motion of particles under the influence of external forces.

The most important argument expounded by Newton against the wave theory of light advanced by Christian Huygens was, however, a very odd observation made and reported by his great opponent (who even honestly admitted to have "found no satisfactory explanation for it.")

What was it? It is well known that a light beam is split by a calcite crystal into an ordinary beam and an extraordinary beam and – provided it is incident orthogonally to the rhombohedral plane – the latter beam is shifted to the side. Then the two beams lie in one plane, the so-called principal intersection of the incident beam.

Huygens arranged vertically two calcite crystals with different orientations and let a light beam impinge from above. He made the following observation: usually both the ordinary and the extraordinary beam leaving the first crystal were split again in the second crystal into two beams, an ordinary and an extraordinary one. Only when the two crystals were oriented either so that the intersections were mutually parallel or mutually orthogonal did just two beams emerge from the second crystal. Whereas in the first case the ordinary beam remained ordinary in the second crystal (the same naturally applied also to the extraordinary beam), in the second case, in contrast, the ordinary beam of the first crystal was converted into the extraordinary beam of the second crystal, and correspondingly the extraordinary beam of the first crystal was converted into the ordinary.

These last two observations surprised Huygens. He wrote (Roditschew and Frankfurt, 1977, p. 43): "It is now amazing why the beams coming from the air and hitting the lower crystal do not split just as the first beam." In the framework of the wave theory of light – note that we are dealing with a scalar theory similar to the theory of sound, where the oscillations are characterized by alternating expansions and compressions of the medium; at that time no one thought of a possible transverse oscillation! – we face a real dilemma: a wave, using modern terminology, is rotationally symmetric with respect to its propagation direction, or, as Newton formulated it, "Pressions or motions, propagated from a shining body through an uniform medium, must be on all sides alike" (Newton, 1730; Roditschew and Frankfurt, 1977, p. 81). There does not seem to be a reason why the ordinary beam leaving the first crystal, for example, should in any way "take notice" of the orientation of the second crystal.

Newton saw a way of explaining the effect using his particle model of light. The rotation symmetry can be broken by assuming the particles not to be rotationally symmetric but rather to have a kind of "orientation mark." The simplest possible picture is the following: the light particles are not balls but cubes with physically distinguishable sides, and the experiment suggests that opposite sides should be considered equivalent. Newton himself is not explicit about the form of the particles, whether they are cubic or block-like, and is satisfied by ascribing to the light particle four sides, the opposites of which are physically equivalent. We thus deal with two pairs of sides, and Newton called one of these pairs the "sides of unusual refraction."

The orientation of the side pairs with respect to the principal intersection of the calcite crystal determined in Newton's opinion, the future of the light particle when it enters the crystal in the following sense: depending on whether one of the sides of the unusual refraction or one of the other sides is turned towards the "coast of unusual refraction" (this means the orientation of the side is orthogonal to the principal intersection plane), the particle undergoes an extraordinary or an ordinary refraction. Newton emphasizes that this property of a light particle is present from the beginning and is not changed by the refraction in the first crystal. The particles remain the same also, and do not alter their orientation in space.

In detail, the observations of Huygens can now be explained as follows (Fig. 2.3(a)): the original beam is a mixture of particles oriented one or the other

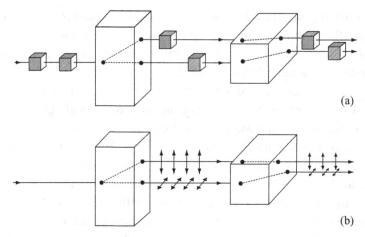

Fig. 2.3.   Passage of a beam through two calcite crystals rotated by 90°. (a) Newton's interpretation; (b) modern description. (The arrows indicate the direction of the electric field strength.)

way with respect to the principal cut plane of the first crystal. The first crystal induces a separation of the particles, depending on their orientation, into an ordinary beam and an extraordinary beam. When the crystals are oriented with their principal intersection planes in parallel, the orientations of the particles with respect to the two crystals are identical. The ordinary beam of the first crystal is also the ordinary beam of the second, and the same applies to the extraordinary beam. However, when the principal intersections of the crystals are mutually orthogonal, the orientation of the particles leaving the first crystal with respect to the second changes so that the ordinary beam becomes the extraordinary beam, and vice versa.

With this penetrating interpretation of Huygens's experiment, in fact Newton succeeded in describing phenomenologically polarization properties of light. Even the name "polarization" was coined by Newon – a fact that is almost forgotten now. (He saw an analogy to the two poles of a magnet.) Today it is well known that the direction to which the "sides of unusual refraction" postulated by Newton points is physically nothing else than the direction of the electric field strength (see Fig. 2.3(b)).

Even though Newton's arguments in favor of a particle nature of light no longer convince us, and the modern concept of photons is supported by completely different experimental facts which Newton could not even divine with his atomic light concept, this ingenious researcher raised an issue which is topical even now. Newton analyzed the simple process of simultaneous reflection and refraction which is observable when a light beam is incident on the surface of a transparent medium. The particle picture describes the process in such a way that a certain percentage of the incident particles is reflected while the rest enters the medium as the refracted beam. In the spirit of deterministic mechanics founded by him, Newton asks what causes a randomly chosen particle to do the one or the other. In fact, the problem is much more acute for us than for Newton because we are now able to perform experiments with individual particles, i.e. photons. While quantum theory sees blind chance at work, Newton postulated the cause of the different behavior to be "fits of easy reflection" and "fits of easy transmission" into which the particles are probably already placed during the process of their emission. These "fits" show a remarkable similarity to the "hidden variables" of the twentieth century which were advocated (as it turned out, unsuccessfully) to overcome the indeterminism of the quantum mechanical description.

We would not be justified, however, in considering Newton, one of the founders of classical optics, to be a blind advocate of the particle theory of light. On the contrary, he was well aware that various observations are understandable only with the aid of a kind of wave concept. He formulated his thoughts in the form of queries with which he supplemented later editions of his *Opticks* (Newton, 1730; Roditschew and Frankfurt, 1977, p. 45), and all of them were characteristically formulated in the grammatical form of negations. It seemed to him that along with the light particles, waves propagating in the "ether" also take part in the game. Is it not so, he asks, that light particles emitted by a source must excite oscillations of the ether when they hit a refracting or reflecting surface, similar to stones being thrown into water? This idea helped him to understand the colors of thin layers which he studied very carefully, mainly in the form of rings formed on soap bubbles (which were named after him).

Altogether we find so many hints for a wave nature of light in Newton's *Opticks* that Thomas Young could cite Newton as "king's evidence" for the wave theory in his lecture "On the theory of light and colours" (Young, 1802; Roditschew and Frankfurt, 1977, p. 153). Even though Young definitely missed the point, it would be just to see Newton as one of the forerunners advocating the dualistic concept that light has a particle as well as a wave aspect (even though he gave emphasis to the first one). From this point of view, Newton seems to us much more modern and nearer in mentality to us than to most of the representatives of nineteenth century physics who were absolutely convinced of the validity of the wave picture.

2.3 Young's interference experiment

> The theory of light and colours, though it did not occupy a large portion of my time, I conceived to be of more importance than all that I have ever done, or ever shall do besides.
>
> *Thomas Young*

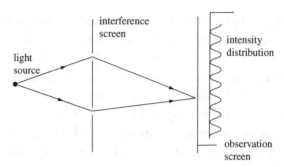

Fig. 2.4. Young's interference experiment.

Interference phenomena are viewed as the most convincing "proof" of the wave nature of light. The pioneering work in this field was carried out by Thomas Young (1773–1829) and, independently, by Augustin Fresnel (1788–1827). It was Young, despite his work going practically unnoticed by his contemporaries, who performed the first interference experiment which found its way into all the textbooks. The principle of the experiment is the spatial superposition of two (we would say coherent) light beams. This is achieved by using an almost point-like monochromatic source and allowing its light to fall onto an opaque screen with two small holes or slits (Fig. 2.4). The two holes themselves become secondary sources of radiation, but because they are excited by the same source they do not radiate independently. We can now position an observation screen at a convenient distance from and parallel to the first screen and we will observe, near the point where the normal (constructed precisely at the center of the straight line connecting the two holes) intersects the observation plane, a system of mutually parallel bright and dark stripes (orthogonal to the aforementioned connecting line), the so-called interference fringes.

The distance (which is always the same) between neighboring stripes depends primarily on the distance between the holes – it is larger when the holes are closer together – and secondly on the color of the light; when we work with sunlight we obtain colored stripes which become superposed at a certain distance from the center of the interference pattern, and the eye gets the impression of a surface uniformly illuminated by white light.

The surprising feature of this observation (let us consider again a monochromatic primary source) is the existence of positions that appear dark even though they are simultaneously illuminated by both holes. When one of the holes is blocked, the interference fringes vanish and the observation screen looks uniformly bright. So, adding light from another hole actually decreases the intensity of the light at some points (so creating the interference fringes).

To stress this point, under certain conditions an equation of the form "light + light = darkness" must hold. However, such a statement is naturally completely incompatible with a particle picture of light as put forward by Newton: when particles are added to those already present, the brightness can only increase. Otherwise the particles would have to "annihilate" each other, which sounds rather odd and contradicts our experience that crossed beams of light mutually penetrate without influencing each other. Thus, the particle picture seems to have been reduced to absurdity.

However, the interference effects are easily explained by the wave picture: when two wave trains (of equal intensity) overlap in certain regions of space, it is certainly possible that the instantaneous "displacement" of one of the waves, compared with the other, is equal in amplitude but opposite in direction (i.e. they are out of phase). In such a case, as is easily demonstrated by water waves, the two displacements balance one another and the medium remains at rest. On the other hand, at positions where the waves oscillate "in phase" (their displacements have the same amplitudes *and* direction) the waves are maximally amplified. Obviously there is a continuous transition between these two limiting cases.

Based on these concepts, Young was able to give a quantitatively correct description of his interference experiment (Young, 1807):

The middle . . . is always light, and the bright stripes at each side are at such distances, that the light coming to them from one of the apertures must have passed through a longer space than that which comes from the other, by an interval which is equal to the breadth[1] of one, two, three or more, of the supposed undulations, while the intervening dark spaces correspond to a difference of half a supposed undulation, of one and a half, of two and a half, or more.

With this interpretation of his observations Young determined the wavelength to be approximately 1/36 000 inch for red light and approximately 1/60 000 inch for violet light, values that agree quite well with known data.

It should be emphasized that the explanation of the interference using the wave concept is independent of the concrete physical mechanism of wave phenomena. In fact, in Young's and Fresnel's days, the physical nature of the oscillatory process taking place was not clear; using the words of the Marquis of Salisbury, scientists were "in search for the nominative of the verb to *undulate*" (to move like a wave). The quest was successfully completed only by the discovery of the electromagnetic nature of light by James Clerk Maxwell (1831–1879).

---

[1] We would say wavelength.

2.4 Einstein's hypothesis of light quanta

> A fundamental change of our concepts of the essence and the constitution of light is indispensable.
>
> *Albert Einstein (1909)*

A pioneering experimental investigation of the photoelectric effect led P. Lenard to surprising results (Lenard, 1902), results that were incompatible with the picture of an electromagnetic wave interacting with an electron bound in some way in a metal. Three years later, Einstein (1905) proposed an interpretation of these experimental facts which openly contradicted the classical light concept. This took the form of a hypothesis on the microscopic structure of light – cautiously declared as "a heuristic point of view concerning the generation and transformation of light", the far reaching consequences of which became obvious only much later in connection with the birth of quantum mechanics.

A thermodynamic approach to the problem of black body radiation led Einstein to the conjecture that light, at least as far as its energy content is concerned, cannot be viewed as if it fills the space continuously; instead it should have a "grainy" structure.

Let us discuss Lenard's observations made when he illuminated a metallic surface (in vacuum) with light and analyzed the electrons released into free space. First of all, he found that the velocity distribution of the electrons is *independent* of the intensity of the incident light.[2] However, the effect was frequency dependent: when a mica or glass plate (absorbing the ultraviolet light component) was inserted in front of the metallic surface, no electrons could be detected. Also, the number of electrons emitted per unit time was proportional to the light intensity, and this held even for very small intensities. In particular, no threshold effect (the onset of electron emission for a certain minimum intensity, an effect almost certainly expected by Lenard) could be observed.

The observation that was least understood was the independence of the kinetic energy of the electrons from the intensity of the light. The expectation was that an electron in the metal should perform, under the influence of an electromagnetic wave, such as light, a kind of resonance oscillation. (Lenard had already discovered that only the ultraviolet part of the spectrum of the used light was of relevance.) During each oscillation the electron absorbs a small fraction of the energy of the wave till the accumulated energy exceeds the potential energy. Then the electron

---

[2] Lenard performed the measurement in such a way that electrons were collected on a metallic disc which was parallel to the illuminated metallic surface. A voltage applied between the two surfaces decelerated the electrons which were emitted in all possible directions from the surface. By varying the applied voltage, it became possible to measure the velocity distribution of the electrons just after their exit from the metallic surface. (More precisely speaking, it is the velocity orthogonal to the surface of the metal that is measured.)

will leave the "potential well" with a certain kinetic energy, and it follows that the energy surplus must have been delivered during the last half or the last full resonance oscillation. Due to this, one should expect that the kinetic energy will be proportional to the intensity of light – in complete contradiction with experience. In addition, for the light intensity used the observed exit velocity was much larger than that predicted by the described model. Under these circumstances, Lenard was forced to look for alternative physical mechanisms of electron emission. He wrote (Lenard, 1902):

Therefore it requires the assumption of more complicated conditions for the motion of the inner parts of the body, but in addition also the, possibly provisional, idea seems to be nearer that the initial velocity of the emitted quanta[3] stems not at all from the light energy but from violent motions already being present within the atoms before the illumination so that the resonance motions play only the role of initiation.

Let us now turn to Einstein's reflections on the subject! This sharp-witted thinker started – similarly to Planck – from thermodynamics considerations. After clarifying that the theoretical foundations of thermodynamics and electrodynamics as applied to the problem of black body radiation fail at small wavelengths and low temperatures of the radiation (and therefore also for low energy densities), he concentrated on the case where Wien's law is still applicable. First he derived an expression for the entropy of a monochromatic radiation field; he was particularly interested in its dependence on the volume of the radiation field. Using this and the fundamental Boltzmann relation $S = k \log W$ (with k being Boltzmann's constant), where S is the entropy and W is, up to a constant factor, the probability, Einstein derived the probability that the radiation field of frequency ν enclosed in a "box" with randomly reflecting walls – due to fluctuations present even in the equilibrium state – *completely* concentrates into a partial volume V_0 of the original box volume V. The expression found in this way was formally identical to that known from the kinetic theory of gases, giving the probability that N molecules limited in their freedom of motion to a volume V will be randomly found in a smaller volume V_0. The ratio between the total energy of the field and the value $h\nu$ (h being Planck's action constant) played the role of the particle number of the electromagnetic field. Einstein consequently came to the following important conclusion: "monochromatic radiation of low density (within the validity range of Wien's radiation formula) behaves with respect to the theory of heat as if it consisted of independent energy quanta of magnitude $h\nu$.[4]" "If this is indeed the case",

---

[3] Lenard meant electrons.

[4] In fact, Einstein, who started from Wien's law not Planck's law, did not use Planck's constant but wrote instead $R\beta/N$, where R is the universal gas constant, N is Avogadro's number and β is the constant in the exponent of Wien's law.

Einstein continued, "it is natural to investigate whether also the laws of generation and transformation of light are of such a kind as if the light would consist of such energy quanta." He imagined that these energy quanta, today called light quanta or photons, are "localized in space points" and "move without being split and can be absorbed or generated only as a whole."

Einstein continued his argument by saying that the new concept of light is appropriate for the understanding of a number of peculiarities found when studying photoluminescence (Stokes's rule), the photoelectric effect and ionization of gases using ultraviolet radiation. Concerning the photoelectric effect we have to adopt the following picture: an incident light quantum immediately transfers all its energy to an electron in the metal. The energy is partly used to release the electron from the metal, i.e. to perform the "exit work" A, while the remaining part is retained by the electron in the form of kinetic energy. Mathematically this relation can be written as

$$h\nu = \frac{1}{2}mv^2 + A, \tag{2.1}$$

where m is the mass of the electron and v is the velocity of the released electron. Because the number of elementary processes will be proportional to the number of incident light quanta, we expect a linear increase of the number of released electrons per second with the light intensity, in agreement with Lenard's observation.

An important consequence of Equation (2.1) is that the kinetic energy of the electrons – for a given material – depends on the frequency, but not on the intensity, of the incident light, and a minimum frequency, the so-called threshold frequency ν_t (determined by the material constant A) must be exceeded to initiate the process. (The predicted linear increase of the kinetic energy with frequency for $\nu > \nu_t$ later allowed the possibility of a very precise and practical method for the measurement of h; more precisely, because the kinetic energy is measured using a compensating electric field, the ratio of Planck's constant and the elementary electric charge e could be determined.)

Einstein finally convinced himself that Equation (2.1) gives the correct (i.e. that found by Lenard) order of magnitude for the kinetic energy – or, on recalculating, for the voltage necessary to decelerate the electrons – when he inserted for the frequency the value of the ultraviolet boundary of the spectrum of the sun (and neglected in first approximation the value of A).

Herewith the observations of Lenard could be considered as "explained." Einstein, however, was much more modest in his formulation: "Our concept and the properties of the light electric effect observed by Mr. Lenard, as far as I can see, are not in contradiction."

It underlines the admirable physical intuition of Einstein that he was also thinking about, as we would say, many-photon processes. In his opinion, deviations from the observed Stokes' rule for fluorescence could be found when "the number of energy quanta per unit volume being simultaneously converted is so large that an energy quantum of the light generated can obtain its energy from several generating quanta." Thus, Einstein was the first to consider the possibility of *non-linear* optics, which, through the development of powerful light sources in the form of lasers, became reality and, with its wealth of effects, became an important discipline in modern physics.

Only in a following paper did Einstein (1906) establish a relationship between the light quantum hypothesis and Planck's theory of black body radiation. He underlined that the physical essence of Planck's hypothesis can be distilled into the following statement: the energy of a material oscillator with eigenfrequency ν_0 interacting with the radiation field can take on only discrete values which are integer multiples of $h\nu_0$; it changes through absorption and emission stepwise in integer multiples of $h\nu_0$.

While the conservative Planck made every effort to reconcile this phenomenon with classical physics – later he confessed (Planck, 1943) that "through several years I tried again and again to build the quantum of action somehow into the system of classical physics" – Einstein took Planck's hypothesis physically seriously. The hypothesis eventually proved to be the first step on the road towards a revolutionary rethinking in physics, which was finalized by the birth of quantum mechanics.

Even though the light quantum hypothesis – in the form of Equation (2.1) – was later confirmed experimentally by the careful measurements of Millikan (1916) (completely against his own expectations!), the question of its compatibility with many optical experiments (such as interference or diffraction), comprehensible only with the concept of waves continuously filling the space, remained open. Einstein was well aware of this and saw the only way out (as he wrote in his paper of 1905) "in relating the optical observations to time averaged values but not to instantaneous values".

However, today this argument is no longer convincing: it has been experimentally confirmed that interference is possible even in situations when at each instant just one photon can be found in the whole apparatus (for instance in a Michelson interferometer) and the photon has to "interfere with itself," as concisely formulated by Dirac (1958). It seems that we cannot avoid also assigning wave properties to individual photons, and Einstein's light quanta hypothesis leads ultimately to a dualistic picture of light.

The following chapters illustrate the ways in which the pioneering work of Einstein was deepened and broadened due to impressive experimental, technical as well as theoretical progress. Before that, let us recall some of the fundamentals of the theory of light based on classical electrodynamics. Classical pictures will guide us through the study of optical phenomena in the case of microscopic elementary processes, and they are, in the end, the criterion for what appears to us "intuitive" and hence "understandable" – or, on the contrary, "paradoxical" or "inconceivable".

3

Basics of the classical description of light

3.1 The electromagnetic field and its energy

The conclusion by Maxwell, based on theoretical considerations, that light is, by its character, an electromagnetic process, is surely a milestone in the history of optics. By formulating the equations bearing his name, Maxwell laid the foundations for the apparently precise description of all optical phenomena. The classical picture of light is characterized by the concept of the electromagnetic field. At each point of space, characterized by a vector **r**, and for each time instant t, we have to imagine vectors describing both the electric and the magnetic field. The time evolution of the field distribution is described by coupled linear partial differential equations: the Maxwell equations.

The electric field strength has a direct physical meaning: if an electrically charged body is placed into the field, it will experience a force given by the product of its charge Q and the electric field strength **E**. (To eliminate a possible distortion of the measured value by the field generated by the probe body itself, its charge should be chosen to be sufficiently small.) Analogously, the magnetic field strength **H**, more precisely the magnetic induction $\mathbf{B} = \mu\mathbf{H}$ (where μ is the permeability), describes the mechanical force acting on a magnetic pole (which is thought of as isolated). Also, the field has an energy content, or, more correctly (because in a precise field theory we can think only about energy being distributed continuously in space), a spatial energy density. The dependence on the field strength can be found in a purely formal way: starting from the Maxwell equations and applying a few mathematical operations, we find the following equation know as the Poynting theorem (see, for example, Sommerfeld (1949)):

$$\frac{\partial}{\partial t}\left(\frac{1}{2}\epsilon\mathbf{E}^2 + \frac{1}{2}\mu\mathbf{H}^2\right) + \mathbf{E}\mathbf{J} + \operatorname{div}\mathbf{S} = 0, \tag{3.1}$$

where $\mathbf{J}$ is the electric current density and the Poynting vector, $\mathbf{S}$, is the abbreviation for the vector product of the electric and magnetic field strengths:

$$\mathbf{S} = \mathbf{E} \times \mathbf{H}. \tag{3.2}$$

To keep the analysis simple, we have assumed a homogenous and isotropic medium with a dielectric constant ϵ and a permeability μ.

Integrating Equation (3.1) over an arbitrary volume V, we find (using Gauss's theorem) the relation

$$\frac{\partial}{\partial t} \int_V \left(\frac{1}{2} \epsilon \mathbf{E}^2 + \frac{1}{2} \mu \mathbf{H}^2 \right) d^3 \mathbf{r} + \int_V \mathbf{E} \mathbf{J} \, d^3 \mathbf{r} + \int_O S_n \, df = 0, \tag{3.3}$$

where O is the volume's surface, df is an element of the surface and S_n is the normal component of the Poynting vector $\mathbf{S}$.

The easiest term to interpret in Equation (3.1) is $\mathbf{E} \mathbf{J}$. It is the work done by the electric field per unit time on an electric current (related to a unit of volume), and is usually realized as heat. Due to this it is natural to interpret Equation (3.3) as an energy balance in the following sense: the rate of change of the electromagnetic energy stored in volume V is caused by the work performed on the electric currents present and by the inflow (or outflow) of the energy from the considered volume. This means that the quantity

$$u = \frac{1}{2} \epsilon \mathbf{E}^2 + \frac{1}{2} \mu \mathbf{H}^2 \tag{3.4}$$

can be interpreted as the density of the electromagnetic field energy (split into an electric and a magnetic part) – in analogy to the deformation energy of an elastic medium – while the Poynting vector represents the energy flow density.

The picture we have formed is that the electromagnetic energy is deposited – in the form of a continuous spatial distribution – in the field. In addition, there is also an energy flow (which plays a fundamental role primarily for radiation processes).

It seems that the mathematical description of the energetic properties in the electromagnetic field is uniquely rooted in the theory. In fact this holds only for the energy density. Equation (3.2) for the energy flow is, in contrast, not unique. From Equation (3.1) it is easy to see that the Poynting vector can be supplemented by an arbitrary divergence-free vector field, i.e. a pure vortex field, without changing the energy balance equation, Equation (3.1). This non-uniqueness in the description of the energy flow caused discussions as to which is the most "sensible" ansatz for the flow, and these did not subside until recent times. The discussion showed that the question is far from trivial because, depending on the situation, one or the other expression seems better suited to "visualize" the situation.

We do not want to go deeper into the problem; we just find it noteworthy that the theory does not give us a unique picture of the energy flow. In fact, we can find completely different representations of the energy flow which are physically equivalent.

What the Poynting vector actually demands from our visualization can be easily illustrated by the example of two orthogonal ("crossed") *static* fields: an electric field and a magnetic field. According to Equation (3.2), energy flows continuously through space but cannot be "caught" because the outflow and inflow from an arbitrary volume element are the same.

An interesting aspect of the continuous distribution of energy in space – a concept found not only in Maxwell's theory, but inherent in any classical field theory – is the possibility of *arbitrarily* diluting the energy. For instance, we can send light into space using a searchlight which will have an angular aperture due to diffraction. Hence, the energy contained in a fixed volume will become smaller and smaller as the light propagates further, and there is no limit for the process of dilution. This is in sharp contrast to what is observable for massive particles. For instance, an electron beam is also diluted, i.e. the *mean* particle density decreases in a similar way to the electromagnetic energy density, but an all-or-nothing principle applies to the measurement: either we find some particles in a given volume or we do not find any. With increasing dilution, latter events become more frequent, and in these cases the volume is absolutely empty. This discrepancy is eliminated if we consider the phenomenon in terms of Einstein's light quanta hypothesis, and therefore also in terms of the quantum theory of light, which ascribes particle as well as wave properties to light.

3.2 Intensity and interference

It is interesting to question what are the optically measurable physical quantities when we view the electromagnetic field from the perspective of optics (this includes visual observations). These quantities obviously *cannot* be the electric or the magnetic field strengths themselves because they are rapidly varying in time, and no observer would be able to follow these high frequency oscillations – the duration of an oscillation is about 10^{-15} second. What really happens during the registration of optical phenomena – in a photographic plate or in the eye – is, at least primarily, the release of an electron from an atomic structure through the action of light. Such a photoelectric effect, as described in more detail in Section 5.2, is described by the time averaged value of the square of the electric field strength[1] at the position of the particular detecting atom, i.e. by the variable

---

[1] Very often the intensity is identified, apart from a normalization factor, with the absolute value of the time averaged Poynting vector (compare with Born and Wolf (1964)). This definition is equivalent to Equation (3.5) for running waves, but it fails in the case of standing waves. Then the time averaged Poynting vector vanishes; nevertheless, a photographic plate is blackened at the positions of the maxima of the electric field strength, an effect known from Lippmann color photography.

called intensity:

$$I(\mathbf{r}, t) = \frac{1}{2} \frac{1}{T} \int\limits_{t-\frac{1}{2}T}^{t+\frac{1}{2}T} \mathbf{E}^2(\mathbf{r}, t') \, dt'. \tag{3.5}$$

The averaging extends over a duration of at least several light periods, and the factor of $1/2$ was introduced for convenience to avoid the presence of factors of 2 (see Equation (3.10)).

To calculate the average it is convenient to decompose the electric field strength, generally written in the form of a Fourier integral,

$$\mathbf{E}(t) = \int\limits_{-\infty}^{\infty} f(v) e^{-2\pi i v t} \, dv, \tag{3.6}$$

into the positive and the negative frequency parts,

$$\mathbf{E}(t) = \mathbf{E}^{(+)}(t) + \mathbf{E}^{(-)}(t), \tag{3.7}$$

where

$$\mathbf{E}^{(+)}(t) = \int\limits_{0}^{\infty} f(v) e^{-2\pi i v t} \, dv \tag{3.8}$$

and

$$\mathbf{E}^{(-)}(t) = \int\limits_{-\infty}^{0} f(v) e^{-2\pi i v t} \, dv = \int\limits_{0}^{\infty} f^*(v) e^{2\pi i v t} \, dv = \mathbf{E}^{(+)*}(t), \tag{3.9}$$

and we note that $f(-v) = f^*(v)$ because $\mathbf{E}$ is real.

When the distribution of frequencies is very narrow compared with the central frequency – we refer to this as quasi-monochromatic light – Equation (3.5) for the intensity reduces to the simple form

$$I(\mathbf{r}, t) = \mathbf{E}^{(-)}(\mathbf{r}, t) \mathbf{E}^{(+)}(\mathbf{r}, t). \tag{3.10}$$

There is a fundamental difference between detection methods in acoustics and optics. Sound waves cause mechanical objects (acoustic resonators, the ear-drum) to oscillate, whereas the detection of optical signals is realized through a process non-linear in the electric field strength; a kind of rectification takes place – the conversion of an alternating electric field of extraordinarily high frequency into a time-constant or slowly varying photocurrent. This is the reason for the fundamental difference in our aesthetic abilities of perception of tone on the one hand and of colors on the other.

A phenomenon that is conditioned mainly by the wave nature of light is the interference effect. From a formal point of view it is possible to say that the reason for the interference is the linearity of Maxwell's equations, implying that the sum of two solutions is again a solution. Physically speaking, this superposition principle means the following: for two incident waves, the electric (magnetic) field strength in the area of their overlap is given by the sum of the electric (magnetic) field strengths of the two waves.

Let us consider the simple case of interference of two linearly polarized (in the same direction) plane waves with slightly different frequencies and propagation directions; for the positive frequency part of the electric field strength of the total field we have

$$\mathbf{E}^{(+)}(\mathbf{r}, t) = A_1 \mathbf{e} \exp\{i(\mathbf{k_1 r} - \omega_1 t + \varphi_1)\} + A_2 \mathbf{e} \exp\{i(\mathbf{k_2 r} - \omega_2 t + \varphi_2)\}, \quad (3.11)$$

where A_j are (real) amplitudes, $\mathbf{e}$ is a unit vector indicating the polarization direction, $\mathbf{k}_j$ is the wave number vector, ω_j is the angular frequency and φ_j is the (constant) phase of the partial waves j $(=1, 2)$.

Equation (3.11) can be rewritten as

$$\mathbf{E}^{(+)}(\mathbf{r}, t) = A_1 \mathbf{e} \exp\{i(\mathbf{k_1 r} - \omega_1 t + \varphi_1)\}[1 + \alpha \exp\{i(\Delta \mathbf{kr} - \Delta \omega t + \Delta \varphi)\}], \quad (3.12)$$

where we have used the abbreviations $\alpha = A_2/A_1$, $\Delta \mathbf{k} = \mathbf{k_2} - \mathbf{k_1}$, $\Delta \omega = \omega_2 - \omega_1$ and $\Delta \varphi = \varphi_2 - \varphi_1$. Equation (3.12) describes a wave process differing from an ideal plane wave in that the amplitude is modulated spatially and temporally.

As already noted in the introduction, the electric field strength is not directly observable in the optical domain. What is in fact possible to see, photograph or register in other ways is the intensity (Equation (3.10)) which can be expressed using Equation (3.11) as

$$I(\mathbf{r}, t) = A_1^2 + A_2^2 + 2A_1 A_2 \cos(\Delta \mathbf{kr} - \Delta \omega t + \Delta \varphi), \quad (3.13)$$

and can also be written, because A_j^2 are the intensities I_j of the waves j $(=1, 2)$, as

$$I(\mathbf{r}, t) = I_1 + I_2 + 2\sqrt{I_1 I_2} \cos(\Delta \mathbf{kr} - \Delta \omega t + \Delta \varphi). \quad (3.14)$$

Obviously, the third term on the right hand side of either Equation (3.13) or Equation (3.14) is responsible for the interference. Light incident on an observation screen is detected by the eye as a system of equidistant bright and dark fringes. The contrast becomes stronger as the difference between the intensities I_1, I_2, becomes smaller. In the special case of $I_1 = I_2$, the intensity in the middle of the dark strips is equal to zero.

We have to note that the described interference pattern shifts in time (it drifts away so to speak) when the frequencies do not match exactly. Only for $\Delta\omega = 0$ is a static – and hence also photographically observable – interference pattern present. In the case $\Delta\omega \neq 0$, the intensity, observed at a fixed point, is modulated by the frequency $\Delta\nu = \Delta\omega/2\pi$. Thus, "beat phenomena" can be observed with the help of a photocell, as will be discussed in more detail in Section 5.2.

3.3 Emission of radiation

One of the most important results discovered (purely mathematically) by Maxwell is that the fundamental equations of electrodynamics allow wave-type solutions. The wave propagation velocity was determined by Maxwell, with the help of his ether theory and a mechanical analogy, to be $c = 1/\sqrt{\epsilon\mu}$. This quantity had already been determined by Weber and Kohlrausch, for the vacuum, through electrical measurements. The surprisingly accurate agreement between the found value and the measurement results of Fizeau for the velocity of light in the vacuum led Maxwell to his conclusions regarding the electromagnetic nature of light (Faraday had already guessed at a connection between light and electricity), but it was Heinrich Hertz who generated electromagnetic waves using an electric method for the first time, thus verifying directly through experiment an essential prediction of the theory.

The simplest model of an emitter of electromagnetic energy into free space, is the so-called Hertzian dipole. It is represented by two spatially separated point charges of opposite sign, one of which is able to move along the line connecting the two charges. Applying an external force to one of the charges causes it to undergo a back and forth motion. This implies a time dependent change of the dipole moment $\mathbf{D} = Q\mathbf{a}$, where Q is the absolute value of the charge and $\mathbf{a}$ is the distance vector pointing from the negative to the positive charge. The time derivative of the dipole moment represents an electric current taking over the role of a source term in Maxwell's equations. The theoretical treatment of the radiation problem leads to simple expressions for the electromagnetic field at larger distances from the source, the so-called far field zone. There the electric and the magnetic fields are determined only by the second order time derivative of the electric dipole moment, i.e. through the acceleration of the moving charge, and it is crucial to note that the value of the field strength at point P and time t is determined by the value of the acceleration at an earlier time by $\Delta t = r/c$ (r is the distance between the source and the point of observation and c is the velocity of light). This effect of "retardation" illustrates the fact that an electromagnetic action propagates with the velocity of light.

In particular it can be shown that **E** and **H** are mutually orthogonal and are also orthogonal to the radius vector having its origin at the light source. In addition the field strengths are proportional to $\sin\theta$, where θ is the angle between the position vector (radiation direction) and the dipole direction. The absolute value of the Poynting vector (pointing in the radiation direction) is

$$S = \frac{1}{16\pi^2\epsilon_0 c^3}\frac{\sin^2\theta}{r^2}\ddot{D}^2, \tag{3.15}$$

where ϵ is the dielectric constant of the vacuum. (See, for example, Sommerfeld (1949).) The $1/r^2$ decrease of the energy flow is easily understood: it follows from this that the energy flow per second through a spherical surface is always the same when the propagation of a certain wavefront is followed through space and time, as is required from the energy conservation law.

The angle dependence in Equation (3.15) is a typical radiation characteristic: a dipole does not radiate at all in its oscillation direction but has maximum emission in the orthogonal direction. Similarly, the sensitivity of energy absorption by an antenna is dependent on the incident angle of the incoming wave. (We have probably all experienced this phenomenon when trying to improve a TV signal by means of repositioning the external aerial.) The physical reason for the behavior of the receiver antenna is obvious. Only the electric field component in the dipole direction is able to start the oscillation of the dipole. The interaction is strongest when the electric field strength coincides in direction with the dipole oscillation, and this applies also for emission. Because the electromagnetic field is transverse, this implies an incident or emitted radiation directed orthogonally to the dipole oscillation.

Let us return to the emission of a Hertzian dipole! Of particular importance is the case of a sinusoidal time dependence of the external driving force. A typical example is a radio transmitter. Under such circumstances, the dipole executes harmonic oscillations, as does the emitted electromagnetic field, which is consequently monochromatic. If we consider the energy, we see that the oscillating dipole is continuously losing energy – we refer to this as "radiation damping", which has an attenuating effect on the moving charge. The lost amount of energy must be continuously compensated for by work done by the driving force on the moving charge to guarantee the stationary, monochromatic emission.

Let us denote the frequency of the field by ν; then we obtain for the time averaged amount of energy emitted per unit time into the spatial angle

$d\Omega = \sin\theta\, d\theta\, d\varphi$ the expression

$$S\, d\Omega = \frac{\pi^2 D_0^2 \nu^4}{2\epsilon_0 c^3} \sin^3\theta\, d\theta\, d\varphi, \tag{3.16}$$

where D_0 is the amplitude of the dipole oscillation.

Equation (3.16) does not hold only for macroscopic antennas emitting radio- and microwaves, but is applicable also to microscopic oscillators like atoms and molecules. We have in fact to conceive (non-resonant) scattering of light on atomic objects, so-called Rayleigh scattering, as follows: the incident radiation induces in the molecules dipole moments oscillating with the light frequency, which in turn radiate not only in the forward direction but, according to Equation (3.16), also sideways (thus bringing about the scattering of light). One consequence of the ν^4 law is demonstrated literally in front of our eyes on sunny days. The sky appears blue because of the stronger scattering of blue light compared with that of red light on the molecules in the air. This explanation, stemming from Lord Rayleigh, is, however, not the whole story. Smoluchowski and Einstein realized that irregularities in the spatial distribution of molecules play an essential role. These irregularities prevent the light scattered to the sides from being completely extinguished through the interference of partial waves from individual scattering centers.

3.4 Spectral decomposition

In contrast to the situation for a radio transmitter, in the Hertz experiment we are dealing with a strongly damped (rapidly attenuated in time) dipole oscillation generated with the aid of a spark inductor. We find such processes also in the atomic domain, for instance in the spontaneous emission case. The simplest model that is applicable to this situation is that of an oscillator, an object able to oscillate with frequency ν, which is put into oscillation through a short excitation (for instance, an electron kick). The emission associated with this process causes an exponential attenuation of the dipole oscillation amplitude. The oscillation comes to rest after a finite time, and a pulse of finite length is emitted. If we observe such a pulse from a fixed position, we see that the electric and the magnetic field strengths – after the wavefront has reached the observer – decrease exponentially. Such a pulse cannot be monochromatic. This is easy to see if we consider the spectral decomposition of the electric field strength as a function of time, i.e. if we write for the positive frequency part

$$E^{(+)}(t) = E_0 e^{-2\pi i \nu_0 t - \frac{\kappa}{2} t} = \int_0^\infty f(\nu) e^{-2\pi i \nu t}\, d\nu, \tag{3.17}$$

where we have assumed, for simplicity, linearly polarized light. E is the electric field strength in the polarization direction.

Next we assume that the field $E(t)$ is zero for $t < 0$. The Fourier theorem applied to Equation (3.17) yields

$$f(v) = \int_0^\infty E^{(+)}(t)e^{2\pi i v t}\,dt, \tag{3.18}$$

and explicitly for our case

$$f(v) = \frac{2E_0}{\kappa + 4\pi i(v_0 - v)}, \tag{3.19}$$

i.e. the radiation has a Lorentz-type frequency distribution

$$|f(v)|^2 = \frac{4|E_0|^2}{\kappa^2 + [4\pi(v_0 - v)]^2}, \tag{3.20}$$

with the halfwidth

$$\Delta v = \frac{\kappa}{2\pi}. \tag{3.21}$$

We speak of an emission "line" of width Δv. Noting that $\Delta t = \kappa^{-1}$ characterizes the duration of the emitted pulse (measured by the intensity dependence at a fixed point), we can rewrite Equation (3.21) in the form

$$\Delta v \cdot \Delta t \approx \frac{1}{2\pi}. \tag{3.22}$$

In this form the relation has a general validity for pulses of finite duration; we must assume, however, that the phase of the electric field does not change significantly within Δt. If this should occur however – there might even be uncontrollable phase jumps, for instance caused by the interaction of the dipole with the environment – the right-hand side of Equation (3.22) can become considerably larger than $1/2\pi$. In general, the relation has to be understood in such a way that it defines the minimum value of the linewidth for a pulse of finite duration.

An interesting physical consequence of Equation (3.22) is that the Fourier decomposition of a pulse in a spectral apparatus leads to partial waves which, due to their smaller frequency widths, are considerably longer than the incident wave. How this is achieved in the spectral apparatus is easily illustrated by the example of a Fabry–Perot interferometer.

The device consists of an "air plate" formed by two silver layers S_1 and S_2 deposited on two parallel glass plates (Fig. 3.1). A beam incident on layer S_1 under a certain angle is split by S_1 into a reflected part and a transmitted part. The

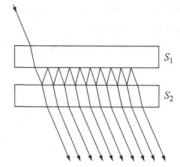

Fig. 3.1.   Path taken by the rays in a Fabry–Perot interferometer (S_1 and S_2 are the silver layers). The rays refracted at S_1 are left out for simplicity.

transmitted beam is incident on S_2; the part of the beam reflected on S_2 is split on the first layer again, and the process continues. The result is the formation of a whole sequence of partial beams that have experienced different numbers of passages between the layers S_1 and S_2. Two neighboring beams differ in their amplitudes by a constant factor, and the path difference Δs between them is also the same and is determined by the geometry of the setup. The superposition of the partial waves yields the total outcoming radiation. Its amplitude reaches its maximum value – this means the transmittivity of the interferometer is maximal – when Δs is an integer multiple of the wavelength of the light. The transmission curve of the Fabry–Perot etalon, as a function of frequency, shows a sequence of maxima with a halfwidth $\delta \nu$ determining the resolution of the apparatus.

For the sake of simplicity, let us assume the spectrum of the incident pulse to be narrow enough to be localized around one of the transmission maxima but to be broad compared with $\delta \nu$; in such a case the interferometer cuts a narrow frequency interval from the spectrum of the incident pulse. According to Equation (3.22) this is associated with a stretching of the pulse. How this happens is now easily understood: each run of the light back and forth between the layers S_1 and S_2 leads to a time delay and – in the case of a pulse – also to a mutual spatial shift of the partial beams: they are increasingly lagging behind the more often they have been reflected between S_1 and S_2.

The spectral decomposition of light is achieved in such a way that the light to be analyzed is incident as a focused beam. Because the path difference Δs depends on the angle of incidence, the photographic image of the outcoming light shows a ring structure, and there is a well defined relation between the ring radius and the light frequency.

The formation of the mutually interfering partial waves requires a finite time δt, which – for high resolution of the spectral apparatus, i.e. for drastic pulse stretching – is essentially equal to the duration of the outcoming pulse. Its linewidth satisfies Equation (3.22), and because it coincides with the transmission width $\Delta\nu$ Equation (3.22) can be interpreted in such a way that $\Delta\nu$ is the precision of the frequency measurement and Δt is the minimum measurement duration. It is not a coincidence that Equation (3.22) in this interpretation represents a special case of the general quantum mechanical uncertainty relation between the precision ΔE of an energy measurement and its duration Δt (Landau and Lifschitz, 1965):

$$\Delta E \cdot \Delta t \geq \frac{h}{2\pi} \equiv \hbar, \tag{3.23}$$

as long as, according to the photon concept, we identify the quantity $h\nu$ (h is Planck's constant) with the energy E of a single photon.

Finally, let us mention that an atom can be "cheated" into experiencing a line broadened radiation field by limiting its interaction time. This occurs, for instance, when the atom passes through a field filling a finite space R. The atom is "seeing" a pulse of finite length and, according to Equation (3.22), in which we have to identify Δt with the time of flight through R, the spectrum appears to the atom to be broadened, even though it might be a monochromatic field. In such cases we speak of the time of flight broadening of an absorption line.

4

Quantum mechanical understanding of light

4.1 Quantum mechanical uncertainty

After reviewing the main characteristics of the classical description of light, let us discuss those aspects of the quantization of the electromagnetic field which are of relevance for the analysis of the phenomena we are interested in. It seems a reasonable place to start to make clear the fundamental difference between the classical and the quantum mechanical description of nature; we will come across this difference many times when discussing experiments, and it will often give us a headache. We have to deal with the physical meaning of what is called uncertainty.

The starting point of the classical description is the conviction that natural processes have a "factual" character. This means that physical variables such as the position or momentum of a particle have, in each single case, a well defined (in general, time dependent) value. However, it will not always be possible to measure all the appropriate variables (for instance, the instantaneous electric field strength of a radiation field); furthermore under normal circumstances we are able to measure only with a finite precision. Hence the basic credo of classical physics should be given in the following form: we are justified in imagining a world with variables possessing well defined values which are not known precisely (or not known at all). In doing this we are not forming any conclusions that contradict our everyday experiences.

This is the fundamental concept of classical statistics: we are satisfied with the use of probability distributions for the variables we are interested in, not from fundamental but purely practical reasons. The necessity is even turned into an advantage in the case of many-particle systems, for instance a gas; what would be the benefit of an absolutely detailed description of, say, 10^{23} particles even if our brain could digest such an overwhelming amount of information? We can make the rather general statement that the uncertainty of classical physics is equivalent to a certain amount of ignorance with respect to an otherwise well defined situation.

A detailed study of the micro-cosmos, however, reveals that the classical reality concept is not applicable. The "factual" must be supplemented with a new category: the "possible." More precisely speaking, in addition to the uncertainty known from classical physics, there is another, specifically quantum mechanical, uncertainty. It is deeply rooted in the foundations of quantum mechanics and it is best known from Heisenberg's uncertainty relation. That the quantum mechanical description of nature is not "intuitive" is most often attributed to this uncertainty.

What is the essence of this quantum mechanical uncertainty? The only precise statement we can make is what the uncertainty is not – namely plain ignorance. Quantum mechanics resorts to a mathematically precise formulation of the new uncertainty concept (which completely satisfies the pragmatist), but it does not provide any clues as to how to form a concrete definition of it. Let us illustrate the problem with a simple example. Consider an ensemble of identical atoms whose energy is not "sharp" in the quantum mechanical sense. (We can prepare such a state by applying a resonant coherent field, which is generated, for instance, by a laser.) Let us assume, for simplicity, that only two levels, 1 and 2, with energies E_1 and E_2 are involved. Quantum mechanics assumes (rightly, as the excellent agreement between prediction and experiment shows) that it is wrong to interpret the aforementioned uncertainty in such a way that a certain percentage of the atoms is in the upper level and the remaining part is in the lower one. Instead we have to assume that there is no difference between the physical states of the individual atoms.[1] The quantum mechanical language for such a situation is the term "pure state." Mathematically, the ensemble of *all* atoms is described by a *single* wave function in the form of a superposition:

$$\Psi = \alpha(t)\Psi_1 + \beta(t)\Psi_2, \qquad (4.1)$$

where Ψ_1 and Ψ_2 describe the eigenfunctions corresponding to the energy levels 1 and 2, and α and β are complex numbers satisfying the normalization condition $|\alpha|^2 + |\beta|^2 = 1$.

When we try to translate Equation (4.1) into plain language, we obtain either "the atoms are in the upper and lower levels *simultaneously*," or "both of the energy levels are possible, but neither of them is a fact."

On the other hand, the specifically quantum mechanical uncertainty is a prerequisite for us to be able to assign to the atoms a coherently oscillating dipole moment (in the sense of a quantum mechanical expectation value) (Paul, 1969). The atomic dipole moment is in a kind of complementary relation to the energy: it vanishes when the atom is in a state of well defined energy. The individual dipole

---

[1] More precisely, we should say: However the ensemble of atoms is split into two ensembles, the atoms always behave in the same way with respect to the measurement of any physical variable (observable).

moments (in the laser, for example, they are induced by the electric field strength residing at the positions of the atoms) add up to a macroscopic polarization oscillating with the frequency of the radiation field, which plays an important role as a source of the laser radiation. (It represents the exact analog of the antenna current of a radio transmitter.)

The transition from the "possible" to the "factual" or from "one *as well as the other*" to "one *or* the other" is realized when an energy measurement is performed, i.e. after a physical interaction. According to the axiomatic description of the quantum mechanical measurement process, a "reduction" of the wave function takes place. It means that the considered ensemble of atoms is turned into an ensemble corresponding to the classical uncertainty concept, which is characterized by a fraction $|\alpha|^2$ of atoms with energy E_1 while the remaining fraction $|\beta|^2$ of atoms has energy E_2. This new ensemble is – in contrast to the original ensemble described by the "pure state" (4.1) – a "statistical mixture" of two sub-ensembles with different atomic energies. A real separation of the two sub-ensembles is obtained by sorting the atoms according to the measured values E_1 and E_2.

As already mentioned, the expectation value of the dipole moments vanishes for such an ensemble. This means that the energy measurement destroys completely the macroscopic polarization of the medium. Physically there is a considerable difference whether we deal with atoms prepared in a pure state (described by the wave function of the form of Equation (4.1)) or with the abovementioned statistical mixture of atoms. The first case is realized in a laser medium (in laser operation), while the second case corresponds to conventional (thermal) light sources. The difference between the two situations is clearly illustrated by the properties of the emitted radiation (for details, see Sections 8.2 and 8.3).

Our example of an ensemble of identical atoms illustrates nicely the disturbing effects of a measurement on a system's dynamics. Let us assume that the atoms are initially in the lower state 1. The intense coherent field induces a transition from 1 to 2, and the theory states that – in the case of resonance – at a certain moment t_A, all the atoms are with certainty in the excited state 2. However, what happens when, from time to time, we check exactly where the individual atoms are, i.e. which level they are in. Our curiosity has dramatic consequences: the corresponding measurement leads to an abrupt interruption of the coherent interaction with the external field, and each of the atoms – according to the probability determined by the wave function at the corresponding instant – is put into either the initial or the excited state. In the first case the interaction has to start from "scratch" while in the latter case an evolution sets in, with the tendency to bring the atoms back into the lower energy state. The massive interruption of the evolution process has the consequence, as the quantum mechanical calculation shows, that at time t_A not all of the atoms by far will be found in the upper state. Moreover, the number of atoms being excited at that time steadily decreases the more frequently the measurement is repeated, until it vanishes completely. Hence, the process of evolution is increasingly hindered the more often we "check." This phenomenon – experimentally verified recently using methods of resonance fluorescence (Itano *et al.*, 1990; see Section 6.1) – resembles the well known Zeno paradox of the flying arrow (which is at each instant located at a well defined position and "consequently" is always at rest) and is often called the quantum Zeno paradox.

Based on the given arguments, we must conclude that we have to take the specifically quantum mechanical uncertainty seriously. Our thinking, which has been conditioned by our preoccupation with classical physics, is completely helpless when faced with the new uncertainty of quantum theory.[2] However, there are certain analogies with classical electrodynamics, because the superposition principle is valid in both theories. For example, the radiation field need not be monochromatic. Generally, the radiation field is a superposition of waves with different frequencies, and the frequency of the wave itself is undetermined in the sense that it can be the frequency of this particular wave as well as that of the other wave in the superposition. Also, the superposition principle plays an important role in the polarization properties of light. Let us consider, for instance, *linearly* polarized light and ask which circular polarization is present. Our answer is that because linearly polarized light can be considered as a superposition of a left handed circularly and a right handed circularly polarized wave, both polarizations are simultaneously present with equal probability.

We should bear in mind, however, that the uncertainty in the classical description of the properties of the electromagnetic field does not contradict the principles of classical physics, which state that all physical processes are "factual." In fact, the electric field strength as the primary physical variable in the above example is well defined with respect to magnitude and direction at each moment, and we arrive at the conclusion that one physical variable has simultaneously different values only when questions of the type: What happens when light is passing through a spectral apparatus? etc. are posed.

We might ask whether a similar situation is present in quantum mechanics. Is the quantum mechanical description in reality incomplete, as was advocated by Einstein, Podolsky and Rosen (1935)? Are its generally statistical predictions simply the result of our ignorance of the exact details of the microscopic processes which we are not able to observe with our (necessarily macroscopic) measurement apparatus? Is it not possible, following the example of classical physics, at least to imagine a more detailed theory from which the statistical predictions of quantum

---

[2] Bohr had already pointed out that we are familiar with such a state of uncertainty from our everyday life: when we face a difficult decision, we have a clear feeling that the different possibilities are in a kind of "simultaneous presence," from which just one is chosen to become real by our will.

mechanics would follow similarly to how statistical physics is derived from classical mechanics?

The possibility of the existence of such a theory, supplemented with additional parameters not accessible to observation and hence called "hidden variables," was denied by the majority of quantum theoreticians. It was Bell (1964) who supplied a convincing proof that this opinion is indeed correct. He managed to show in a surprisingly simple way that there are certain experiments for which quantum mechanical predictions will be distinctly different from predictions of any deterministic hidden variables theory – at least when we exclude the possibility that physical action can propagate faster than light (for details, see Section 11.3).

Hence, there is no hope that the specifically quantum mechanical uncertainty can be done away with by a further development of the theory, however clever it might be. Rather, it shows a qualitatively new aspect of reality.

Let us now turn to the concrete predictions of quantum theory of radiation which are of particular interest to optics.

4.2 Quantization of electromagnetic energy

Fortunately, it is possible to apply without great difficulty the quantization methods for material systems to the radiation field if we consider the eigenoscillations, or modes, of the field. The eigenoscillations emerge quite naturally when we consider a field confined in a resonator with ideally reflecting walls. Physical boundary conditions guarantee that only certain spatial field distributions can form – the infinite conductivity of the resonator wall material forces the tangential components of the electric field strength to vanish, i.e. the field must have nodes at the resonator walls, – and such eigenoscillations oscillate with discrete frequencies (called eigenfrequencies) determined by the geometry of the resonator. The excitation of an eigenoscillation is described by a phase and an amplitude, both of which can be cast into a complex amplitude. Any arbitrary excitation state is generally given by a superposition of such oscillations. This type of "mode picture" is convenient for the quantum mechanical description of the field because each mode can, by itself, be quantized. The individual modes correspond to independent degrees of freedom of the field; in particular, the field energy is given as the (additive) sum of the energies of each of the modes.

From a physical point of view, however, the picture of a field enclosed in a "box" is realistic only for the case of microwaves. In optics we deal primarily with free fields which propagate undisturbed – hence they are running waves and not standing waves – and are only altered later through optical elements or by other interaction with matter. To be able to use the convenient mode picture we resort to a trick: we apply fictitious boundary conditions to the free field. We require the field to be periodic in all three spatial directions with a (large) period length L. The aforementioned resonator volume is replaced by a "periodicity cube" with an edge length L. The modes selected by the boundary conditions are monochromatic running plane waves differing in propagation direction, frequency and polarization. (The end points of the wave number vectors form a cubic lattice with a lattice constant $2\pi/L$.) Because the length L does not have any physical meaning, we can eliminate it by calculating first with a fixed L value and then, once we have the final result, performing the limit $L \to \infty$.

What is accomplished by this procedure is the description of the radiation field by an infinite *countable* manifold of degrees of freedom. This is realized by the selection of discrete modes out of the actually present continuum of modes (represented by a continuous distribution of wave vectors).

Our task now is to describe quantum mechanically a single radiation mode. It turns out that this problem has already been completely solved quantum mechanically for material systems; the results can simply be taken over. A single mode of the radiation field is formally identical to a single (one-dimensional) harmonic oscillator. This is easily seen because a field excited in a single eigenoscillation will oscillate harmonically. This means that the time dependent (complex) amplitude has the form

$$A(t) = A_0 \exp(-i\omega t) = |A_0| \exp\{-i(\omega t - \varphi)\},$$

and its real part $x = |A_0| \cos(\omega t - \varphi)$ satisfies the equation

$$\ddot{x} + \omega^2 x = 0. \tag{4.2}$$

This is simply the equation of motion of a harmonic oscillator, with x being the displacement of the particle from the equilibrium position. Let us note that the radiation mode is the only known *exact* realization of a harmonic oscillator; the mechanical harmonic oscillator – based on the validity of Hooke's law of a restoring force proportional to the displacement – can be viewed as an approximation valid only for small displacements.

The fundamental result of the quantum mechanical description of the harmonic oscillator is that its energy is quantized in the form of equidistant steps (Fig. 4.1). The same applies also to the energy of the radiation mode which we have to envisage as being distributed over the whole mode volume according to the spatial distribution of the field intensity.

Counting the energy from the lowest level, the energy E_n of the nth excitation state is an nth multiple of a constant, $h\nu$:

$$E_n = nh\nu \qquad (n = 0, 1, 2, \ldots) \tag{4.3}$$

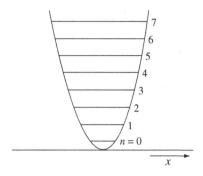

Fig. 4.1. Energy levels of a harmonic oscillator. The parabolic curve represents the potential energy as a function of the displacement x.

(see Section 15.1). We can say that when the quantum mechanical state with energy E_n is realized, n "energy packets" of magnitude $h\nu$ (photons) exist in the mode volume. According to the different field modes, there are different sorts of photons. Hence, a state of sharp energy of the total field is characterized by an infinite set of photon numbers which are related to the particular modes. When the photon number of one mode equals zero, we say that the mode is in the vacuum state.

We should comment on the zero point energy of the electromagnetic field. In fact, the ground state of the harmonic oscillator is associated with the energy $\frac{1}{2}h\nu$, whereby the zero point of the energy scale was chosen in such a way that the potential energy of the elastically bound particle in its equilibrium position is zero (Fig. 4.1). Adding this energy to the energy of the radiation field, we find that the total energy (the sum of all mode contributions) is hopelessly divergent. This is, however, not too disturbing because the energy cannot be exploited in any way. Were it possible to "tap" a part of this energy, then the whole field would obviously be transferred into a state with lower energy; such a state cannot exist, however, because the vacuum state is, by definition, the lowest energy state of the field. It seems natural to dispose of the zero point energy of the electromagnetic field simply by making it the new zero of the energy scale (as we did above).

There is a situation, however, where the argument is not so simple. Let us consider the case of a real resonator with a variable form – in the following we will focus on the space between two plane parallel plates separated by a very small distance z – then the zero point energy of the enclosed field diverges, but it also changes with the separation length z by a finite amount, and the above "renormalization" of the field energy can be achieved only for a single value of z. In fact, the metallic plates cause an upper cutoff in the wavelength spectrum of the allowed modes. Because the electric field strength must have nodes (points of zero value) on the metallic surfaces, the greatest possible value of the wavelength is z. With increasing z, more modes are available, and hence the zero point energy increases.

Interpreting the zero point energy as the potential energy $V(z)$ of the system, we can derive a force **K** using the well known formula **K** $= -$**grad** V. This force is orthogonal to the plates and is negative because $dV/dz > 0$, i.e. it describes an attraction of the plates. The force was theoretically predicted by Casimir (Power, 1964), who derived the following value for it:

$$K = -\frac{\pi}{480}\frac{hc}{z^4},$$
(4.4)

(where c is the velocity of light). The formula reveals that the Casimir force is a pure "quantum force" between electrically neutral bodies. It is proportional to Planck's quantum of action, and hence has no classical analog. It has been measured precisely more recently by Arnold, Hunklinger and Dransfeld (1979). In fact, it is simply a special case of the well known Van der Waals forces which generally act between macroscopic bodies (with a very small separation) whose surfaces form something like (partially open) resonators. The forces are very weak and of very short range. In addition, their strength depends on the dielectric properties of the respective material.

Let us return to the radiation modes. These are physically defined only when we deal with resonator modes; they are individually excitable and the mode volume equals the resonator volume. Due to this, the (spatial) photon density, which is of paramount importance for the interaction with matter, is non-zero for finite photon numbers. This, however, does not apply to the radiation modes introduced with the aid of artificial periodic conditions which have the whole space at their disposal. In this case we should identify the mode volume with the fictitious, in principle infinitely large, periodicity cube. It is obvious that such individual modes cannot be excited. This is not possible for the simple reason that we would need an infinite number of photons (i.e. infinite energy) to obtain a non-vanishing photon density. The situation is basically no different from the one in classical electrodynamics, where we also find that an ideal plane wave is not realizable. Realistic fields with a necessarily finite linewidth and a finite spatial (as well as temporal) extension are, from the classical viewpoint, superpositions of infinitely many plane waves. Quantum mechanically they are represented in the simplest case by energy eigenstates of the field; the photon numbers are non-zero for those modes which are compatible with the parameters of the light beam (mean frequency and linewidth, propagation direction and its uncertainty, and polarization). Then, the mean photon number per mode obtained by averaging over all the excited modes indeed has a well defined finite value that is uniquely determined by the energy or photon density related to unit volume and frequency interval and is independent of the mode volume V (assumed to be large). The reason is that the increase of the volume leads to a linear increase in the density of modes, and hence the number of excited modes increases linearly with V.

The preceding description of a realistic electromagnetic field is rather laborious. It would be convenient if the resonator mode approach could also be used in the case of realistic fields. This is indeed the case when we develop a realistic mode concept. To do so, let us recall that the characteristic of a mode is that it represents a system with a single degree of freedom. Such a requirement is obviously satisfied by a coherent pulse with a prescribed shape.[3] What still remains open to question are the concrete values of the amplitude and the phase of the pulse, which are usually cast into the form of a complex amplitude that represents, just as in the case of a resonator mode or modes defined by periodicity conditions, the dynamical variables of the system which change due to interactions. On the other hand, in the case of continuous (usually stationary) irradiation by an almost monochromatic (so-called quasimonochromatic) light, we can identify the mode with that part of the light beam that fills a volume of the size of the coherence volume. The coherence volume may be understood as a part of the space (filled with the light field) with a length equal to the longitudinal coherence length in the radiation direction and equal to the transverse coherence length in the orthogonal direction. The spatial change of the instantaneous amplitude and phase in such a volume is very small by definition, and the discussed variables remain constant during the undisturbed time evolution (observed from a co-moving frame), i.e. during propagation, and so the single mode formalism is applicable at least approximately.

This realistic mode formalism works for running plane waves, but we must keep in mind that the description applies only to finite space-time regions. In fact, we follow the path used extensively in classical optics (Sommerfeld, 1950). A central role in the quantum mechanical description of light is played by the concept of the photon number *related to the mode volume* and so, for the single mode description, a physical definition of this volume is indispensable, and the most important result of the previous analysis is that we have actually given such a definition.

A considerable difference exists between the photon concept developed on a quantum mechanical basis and that described by Einstein: whereas Einstein conceived photons as spatially localized particles, the quantum mechanical photon concept is related to macroscopic volume. As in the classical description, the energy is spatially distributed. The "photons of a theoretician," as we might call them, are not at all the same thing as those described by an experimentalist who reports that a photon was absorbed at a particular position or was "registered" by a detector. In fact, the grainy structure of the electromagnetic field Einstein had in

---

[3] Experimentally, we need a sequence of nearly identical pulses, typically generated by pico- and femtosecond lasers.

mind appears only when an interaction with (localized) particles takes place. We will discuss this problem in detail later (Section 5.3).

4.3 Fluctuations of the electromagnetic field

The states of the electromagnetic field oscillating in a single eigenoscillation, with a sharp energy, mentioned in the previous subsection, are numbered by the index n, which we can interpret as the photon number. We can say equally well that we are dealing with eigenstates $|n\rangle$ of the photon number operator. The wave properties of the field are determined by the electric field strength. In the quantum mechanical description, the field is described by a Hermitian operator. For the description of optical phenomena the formal fact that the photon number operator and the operator of the electric field strength[4] do not commute is of particular importance (see Section 15.1). The consequence is that for states with a sharp photon number the electric field strength is necessarily indeterminate – in the quantum mechanical sense explained in Section 4.1. In particular, the quantum mechanical expectation value of the electric field strength at any time is zero (see Section 15.1). This does not however imply that the electric field strength measured at time t on an ensemble of radiation fields characterized by the wave function $|n\rangle$ is always zero. The situation is such that the (instantaneous) electric field strength fluctuates around the zero value, and positive as well as negative (with the same modulus) measurement values appear with the same probability. Because the measurement time is fixed, these fluctuations have nothing to do with the time evolution; they just indicate a complete uncertainty of the phase of the electric field strength. Because the system is in a pure state, this uncertainty must not be confused with simple ignorance, rather it is based on fundamental reasons, as explained in Section 4.1. Of particular interest for us is the case $n = 1$: according to what has been said above the phase of a single photon must be viewed as fundamentally uncertain.

The electric field strength does not come to rest, even in the lowest energy state where the photon number zero corresponds to the vacuum state $|0\rangle$. In this case we speak of vacuum fluctuations of the electromagnetic field. This fact necessarily follows from the quantization of the radiation field (see Section 15.2), and is the exact analog of the zero point oscillations of the harmonic oscillator which also occur, according to quantum mechanics, in the ground state. The particle cannot rest at the lowest point of the potential well as it does in the classical case because then, according to Heisenberg's uncertainty relation for position and momentum, the uncertainty in the momentum would be infinitely large.

---

[4] Usually the magnetic field strength does not play a role in optics.

Quantum theory forces us to revise drastically our concept of the electromagnetic field which was previously formed by classical electrodynamics. Whereas in the classical description the electromagnetic energy density (see Equation (3.4)), and hence also the energy stored in an arbitrary volume, is uniquely determined by the electric and magnetic field strengths so that the energy is sharp, quantum theory declares that the energy is sharp only at the cost of an uncertainty in the field strengths. Conversely, a sharp value of the electric field strength requires, according to quantum theory, an uncertainty in the energy and hence the photon number. Quantum theory, so to speak, loosens the "rigid connection" between electric and magnetic field strengths and energy, postulated by the classical theory, in favor of a more flexible relation.

The discrepancy between the classical and the quantum mechanical descriptions is particularly striking for the vacuum state. In the classical picture the space is completely field free, i.e. actually empty, while quantum mechanics assigns to it a fluctuating electromagnetic field. A measurement of the (instantaneous) electric field strength (there is no known way of performing such a measurement, but no one can forbid a Gedanken experiment) would detect in general, according to quantum mechanics, non-zero values even when no photons are present with certainty. The vacuum fluctuations are not connected to electromagnetic energy, at least not to a usable one.

4.4 Coherent states of the radiation field

Even though quantum mechanics forbids us to assign to an electromagnetic wave sharp values of both the photon number and the phase, we can certainly ask which quantum mechanical state comes closest to representing the classical waves of well defined energy (respectively amplitude) and phase. A satisfactory answer can be obtained with the requirement that the electric field strength should – averaged over time and for a fixed mean photon number – fluctuate as little as possible (see Section 15.2). In this way we arrive (uniquely) at the so-called coherent states of the electromagnetic field, often called Glauber states after the American scientist who was the first to realize their extraordinary usefulness for the description of optical phenomena. The explicit form of the states in question is

$$|\alpha\rangle = \sum_{n=0}^{\infty} e^{-\frac{|\alpha|^2}{2}} \frac{\alpha^n}{\sqrt{n!}} |n\rangle. \tag{4.5}$$

Similarly to the states $|n\rangle$ they describe the excitation of a single mode of the radiation field.

The arbitrary complex number α in Equation (4.5) corresponds to the complex amplitude of the wave (in proper normalization). The modulus squared of α gives us the mean photon number, and the phase of α is the phase of the electromagnetic field, or, more precisely, its quantum mechanical expectation value. The electric field strength fluctuates also in coherent states, which results in a phase uncertainty, the importance of which, however, decreases with increasing mean photon number.

The photon number also fluctuates in the coherent state, as can easily be seen from Equation (4.5). The squared moduli p_n of the expansion coefficients giving the probability of finding, using photon counting, exactly n photons follow a Poisson distribution:

$$p_n = \mathrm{e}^{-|\alpha|^2} \frac{|\alpha|^{2n}}{n!} \tag{4.6}$$

(see Fig. 8.6). As can easily be shown, the variance of the photon number $\Delta n^2 \equiv (n - \bar{n})^2$ (the bar stands for the average) is, for a Poisson distribution, equal to $\bar{n}$.

The coherent states of the electromagnetic field are distinguished by a balanced relation between the fluctuations of the phase on the one hand and the fluctuations of the photon number on the other, hence coming as close as possible to the classical ideal of a sharply defined phase and amplitude without actually reaching it. Only in the limit of an infinitely large mean photon number are the uncertainties in phase and photon number without importance. (From what has been said above it follows that the relative variance $\Delta n^2/\bar{n}^2$ behaves for $\bar{n} \to \infty$ as $1/\bar{n}$ and so goes to zero.) The result proves for this case the validity of Bohr's correspondence principle.

Of great practical importance is the fact that with the laser we have an instrument at our disposal with which we can produce Glauber states. The laser radiation is by itself a very good approximation to a coherent state because the laser process has an amplitude stabilizing effect (see Section 8.3). Of equal importance for quantum optics is the fact that the properties of laser radiation are preserved when it is attenuated by (one photon) absorption. In fact, it was shown early on (Brunner, Paul and Richter, 1964, 1965) that Glauber states remain Glauber states during the process of damping (see Section 15.4). What is changed is merely the value of α.

After this short visit into "gray theory," let us return to our main aim, which is to understand the photon as part of our physical experience. Because our physical experiences can only be quantified by measurement (including evaluation using our senses) it is appropriate to turn our attention to the problem of the experimental detection of light.

5

Light detectors

5.1 Light absorption

Whereas receiving radio waves is a macroscopic process and hence belongs to the area of classical electrodynamics – in a macroscopic antenna an electric voltage is induced whereby a large number of electrons follow the electric field strength of the incident wave, in a kind of collective motion – the detection of light, so far as the elementary process is concerned, takes place in microscopic type objects such as atoms and molecules. As a consequence, the response of an optical detector is determined by the microstructure of matter. In particular, it is impossible – due to the enormously high frequency of light (in the region of 10^{15} Hz) – to measure the electric field strength. What is in fact detectable is the energy transfer from the radiation field to the atomic receiver, and this allows us to draw conclusions about the (instantaneous) intensity of light.

We might ask what we can say about the above-mentioned absorption process from an experimentalist's point of view. Among the basic experiences that provide an insight into the structure of the micro-cosmos is the resonance character of the interaction between light and an atomic system. The atomic system, when hit by light, behaves like a resonator with certain resonance frequencies; i.e. it becomes excited (takes up energy) only when the light frequency coincides with a value that is characteristic for the particular atom. Hence, an incident light wave with an initial broadband frequency spectrum that has passed through a gas exhibits in its spectrum dark zones, the so-called absorption lines. This experimental fact was discovered for the first time by Fraunhofer for sunlight and it forms the basis of absorption spectroscopy which helps to detect reliably the smallest amounts of substances.

The resonator characteristics of the atoms are also evident in the process of emission: the frequency spectrum of the light emitted by excited atoms is built up from discrete "lines" matching exactly the absorption lines.

The described observations present the atom as an object that is capable of oscillating with characteristic frequencies which show up in both absorption and emission. Confining ourselves to just one frequency, we find the atom to be similar to a Hertzian dipole, which, in the resonance case, absorbs energy from the field (for a properly chosen phase relation between the dipole and the external field) or emits energy when the external field is absent. A detailed study of the atomic excitation process reveals another rather peculiar aspect. In a famous experiment, Franck and Hertz (1913, 1914) found that an electron beam is able to transfer energy to atoms only when the kinetic energy is equal to or exceeds a certain minimal value. It soon became clear that the result fitted neatly into the atom model developed by Niels Bohr in 1913. The first Bohr postulate assumed the existence of "stationary" states of an atomic system with certain, discrete, energies. This implies that it is necessary to supply the atom with a certain amount of energy (corresponding to the distance to one of the higher energy levels) in order to alter its state.

This is a peculiarity of microscopic systems, having no counterpart in the classical world, which made a radical departure from classical physics unavoidable. Within the framework of classical physics it is really not possible to understand why an electron bound in an atom should be "allowed" to move only in certain orbits. The explanation for the existence of such stationary states was one of the main objectives of the later developed quantum mechanics.

We might ask how two so different experimental facts are related: on the one hand, the resonance behavior of the atoms in the interaction with light, and, on the other hand, the structure of atoms that manifests itself in the discrete energy levels. The answer (which was later explained in detail by quantum mechanics) was given by Bohr in his second postulate, which stated that the atom can execute jump-like transitions from one level to another associated with the emission or absorption of light. Depending on whether the transition takes place from a higher energy level to a lower energy level, or vice versa, we have emission or absorption. The frequency of the corresponding spectral line is

$$\nu = \frac{1}{h}(E_m - E_n). \tag{5.1}$$

Here, E_m is the energy of the upper level, E_n is the energy of the lower level and h is Planck's constant. The equation constitutes a unique relation between the energy levels and the atomic resonance frequencies.

Actually, the fundamental relation Equation (5.1) had already been directly verified by Franck and Hertz, who measured the excitation energy of mercury atoms using the electron collision method and the frequency of the fluorescence light produced.

However, the process of absorption by an atomic system is by no means a measurement of light. Quite generally, we may only speak of a physical measurement when it leads to a macroscopically fixed result (for instance in the form of a pointer deflection); the observed object has to induce in the final stage an irreversible macroscopic change.

For an absorber such a process is the conversion of the excitation energy of the individual atoms or molecules into thermal energy. (In the case of a gas, particle collisions ensure the conversion of the excitation energy into kinetic energy.) The result is a temperature increase, which is detectable with conventional methods.

Fortunately, the inclusion of the rather "inert" thermodynamic processes into the measurement can be circumvented – in favor of electric or electrochemical processes – by using the photoelectric effect, in which the incident light causes not excitation but ionization of the atoms. The elementary step of the process, the release of an electron from the atom, is by itself an irreversible process. The electron leaves the atom due to its acquired kinetic energy, and there is virtually no chance that it will be recaptured. The situation is very similar to that of spontaneous emission (see Section 6.5). The necessarily macroscopic measurement process then requires only an appropriate amplification of the microscopic primary signal.

5.2 Photoelectric detection of light

We present first in this section a chronological discussion of the main types of photoelectric receivers.

The first practical application of the photoeffect was photography, in which the primary process is the release of a valence electron from one of the bromine ions in a silver bromide grain (taking the form of an ionic crystal Ag^+Br^-) initiated by incident light. The electron moves freely in the crystal lattice (hence we speak of an inner photoeffect) and can be captured by one of the crystal defects. When this happens, the defect becomes negatively charged and is able to attract one of the silver ions sitting in the interstitial sites (which explains why it can move freely) and neutralize its charge. This process can be repeated several times at the same defect by a repeated capture of electrons. The result is the formation of a "nucleus" consisting of several silver atoms. It is the development, a chemical treatment of the photographic layer in which a grain containing such a nucleus is reduced to a black, metallic silver grain as a whole, that makes the photographic picture visible, thus producing a macroscopic "trace" of the light involved.

One significant drawback of the photographic procedure is that the long exposure times required do not allow us to obtain any information about the behavior of light quickly enough. It is possible, however, to measure brief changes of intensity over time by using the photoelectrically released electron directly in the detection process. This principle is realized in the form of a photocell, where light releases electrons from an appropriately coated metallic surface. The electrons are emitted into the vacuum and are guided with the help of an externally applied voltage to an anode. The strength of the resulting electric current in the external circuit is a measure of the intensity of the light incident on the photocathode.

To be more precise, there is (within the classical description valid for intensities that are not too small) a relation between the electric current J and the light intensity I of the form

$$J(t) = \alpha \frac{1}{T_1} \int_{t-T_1}^{t} I(t')\, dt', \qquad (5.2)$$

where α is the sensitivity of the device. The time averaging in Equation (5.2) has its origin mainly in the fact that a single released electron generates a current pulse whose duration equals the passage time of the electron between the cathode and the anode. The minimal value of T_1 obtainable in practice – determining the resolution time of the photocell – is about 10^{-10} s. Thus, the photocell is able to follow intensity changes that take place on a time scale $\Delta t \geq T_1$. Correspondingly, the photocurrent is not constant in time but contains a direct current component as well as an alternating current component at frequencies not exceeding the value $1/T_1$.

The creation of the photocell opened up to researchers a new dimension of experimental investigation: for the first time they had in their hands an instrument allowing them to observe directly the fluctuation phenomena of light and hence its microscopic structure.

The detection efficiency of the photocell, however, is far from ideal. Obviously many electrons are needed to obtain a measurable current, hence the intensity of the incident light must not be too low. Important progress was achieved by the development of the secondary emission multiplier, or photomultiplier. The principle of the device is that each released electron becomes the "ancestor" of an avalanche of electrons. The details of the process are as follows. The primary electrons are accelerated by an electric field and impinge onto the first auxiliary electrode (a so-called dynode), releasing a group of secondary electrons. These are again accelerated and hit the second dynode, where they generate new "offspring" and so on. The procedure is repeated several times till the generated electron avalanche is sucked up by the cathode.

The photomultiplier not only allows the conversion of very weak light fluxes into electric currents, it also makes possible the detection of individual photons, thus reaching the utmost frontier of light registration set up by Nature through the atomistic structure of matter. A single primary electron causes an electron avalanche, which induces a current pulse in an external circuit, which is transformed into an easily measurable voltage pulse on an Ohmic resistor, which represents the macroscopic signal indicating the detection of a single photon. We call such a device a photodetector or photocounter.

We should point out that the visual process taking place in our eye relies also on the principle of a photodetector. Actually, Nature equipped us with the most sensitive detection instrument available: it is a proven that an adapted eye is able to "see" just a few photons!

Not all of the photons incident on the cathode induce a photoeffect (at least in the visible spectrum, though in the ultraviolet part of the spectrum the requirement that each incident photon generates an electron can be satisfied), and hence an incident photon will not be detected with certainty but only with a certain finite probability. Thus, the counter sensitivity, which depends primarily on the cathode material and the spectral range, takes different values.

The probability of a photoelectric release of an electron is proportional to the instantaneous light intensity, though, as with the photocell, passage time effects will limit the resolution of the apparatus. Hence, we expect the response probability W of a photodector (per unit time) – in analogy to Equation (5.2) which is valid for the photocell – to be of the form

$$W(t) = \beta \frac{1}{T_1} \int_{t-T_1}^{t} I(t') \, dt'. \tag{5.3}$$

The value of the constant β is determined by the counter sensitivity. The integration time T_1, the so-called response time of the detector, is approximately of the same order of magnitude as for the photocell.

Equation (5.3) is based on a classical description. However, this is no longer justified when we enter the field of very small intensities. Equation (5.3) is in flagrant contradiction to experience when applied to wave packets with an energy smaller than $h\nu$ – classically a completely legitimate assumption – because even then it yields a non-zero response probability. However, in such a case photoionization is energetically impossible!

A helping hand is offered by the fully quantized theory (see, for example, Glauber (1965)), replacing Equation (5.3) by

$$W(t) = \beta \frac{1}{T_1} \int_{t-T_1}^{t} \langle \hat{\mathbf{E}}^{(-)}(t') \hat{\mathbf{E}}^{(+)}(t') \rangle \, dt', \tag{5.4}$$

where $\hat{\mathbf{E}}^{(-)}$ and $\hat{\mathbf{E}}^{(+)}$ represent the operators of the negative and the positive frequency parts, respectively, of the electric field strength (see Equations (3.8) and (3.9)). The quantum mechanical expectation value in the integral depends crucially on the ordering of the factors. The form given is the so-called "normal ordering" (see Section 15.1), which is distinguished by the fact that corresponding expectation values are free from vacuum contributions. The vacuum fluctuations of the electric field described in Section 4.3 thus do not have, fortunately, any influence on the photoelectric detection process; in particular, there is no response from the photodetector when (with certainty) no photons are present.

Photoelectrons can also be detected optically. For this purpose a photoelectric image converter is used in which electrons released from a photocathode by the incident light are, after being accelerated by a static electric field, incident on a fluorescence screen. (The whole process takes place in a vacuum tube.) Electron-optical imaging of the cathode onto the screen makes it possible to visualize the intensity distribution of the incident light on the cathode. In this way it is possible to amplify considerably the brightness of the primary (optical) image and, in addition, to adjust the scale of imaging.

Recently, the method of photoelectric image conversion was applied with great success to the measurement of the time dependence of extremely short optical signals. In this case, the troubling influence of electrons on the anode is absent, and hence the time resolution is no longer limited by the time of flight of electrons in the vacuum tube but by the fact that electrons released in different depths of the cathode layer reach the surface at different times (they are decelerated and scattered by interaction with the cathode material). With the help of so-called "streak cameras," a resolution time of one picosecond has been achieved. The electron beam is spatially resolved by applying as linearly as possible a rapidly increasing electric field orthogonal to the propagation direction of the beam, so that the temporal sequence of electrons in the beam is translated into spatial separation on the fluorescent screen.

Currently semiconductor materials are utilized more and more for photoelectric detection. The elementary process is in this case the generation of an electron–hole pair by an incident photon, i.e. an inner photoeffect. So-called photodiodes are used most commonly. These can be combined in large numbers to form single array or matrix configurations and so allow a spatial resolution of the optical signal. We speak in such cases of photodiode arrays. Because photodiodes allow a large amplification via internal avalanche processes, such detectors are called avalanche photodiodes. In the following, we list the possible experiments realizable with the detectors discussed in this section.

Mean intensity measurement

Mean intensity measurements can be carried out with the help of the photographic plate, the photocell and the photomultiplier. The blackening of a photographic plate is a measure of the light intensity averaged over the exposure time. In photo-electric detection the corresponding information is contained in the direct current part of the photocurrent. Using a photodetector we sum up all the events for a longer period of time, and thus we are able to detect even extremely small light intensities.

When we do not confine ourselves to the detection of the intensity at a particular position but instead detect the intensity at different points, we can observe a spatial interference pattern.

Detection of beats

The time analog of spatial interference, the beats discussed in Section 3.2, can be directly observed by letting two coherent light beams with slightly different fre-quencies (delivered in an almost ideal form by two gas lasers for instance) impinge on a photocell or a photomultiplier. They are represented by the alternating current component of the photocurrent, and can be measured accurately by high-frequency engineering. These experimental methods are of considerable practical importance for the high precision measurement of light frequencies, allowing, for example, the measurement of the frequency of a laser with respect to the frequency of a second laser with extreme precision.

Measurement of intensity correlations

We can measure spatial intensity correlations by placing two (or more) detectors at different positions and counting coincidences; i.e. we register only those events for which all the detectors respond at the same time. Such measurements can also be made at higher intensities using photocells (or photomultipliers). The correlations between the photocurrents of the two photocells are determined with the help of a "correlator" (for details, see Section 8.1).

Analogously, temporal intensity correlations (at a fixed position) can be mea-sured. The experimental difficulty of positioning two detectors at the same place is solved in an elegant way following the procedure of R. Hanbury Brown and R. Q. Twiss (1956a), in which the light to be analyzed is split by a beamsplitter (semitransparent mirror) into mutually coherent parts incident on two independent detectors (see Fig. 8.3). The measured variable is then the number of delayed coin-cidences; i.e. only those events are counted in which a response of the first detector after a time delay of τ seconds is followed by a response of the second detector. This experiment can also be carried out for sufficient intensity using photocells and photomultipliers.

Single-photon arrival time measurement

For weak intensities of the incident light (when necessary it can be further attenuated) the photodetector can be used to observe the time "points" at which a photoelectron was released, i.e. moments at which a photon was actually absorbed.
Such measurement techniques can answer questions of the type: How long, after
the onset of irradiation, does it take before the first photoelectron shows up? The
measurement data can also be used to determine the frequency with which two
photons with a prescribed time delay τ are registered. The obtained information
is the same as in the case of delayed coincidence measurements. The practical
difference between the single-photon arrival time measurement method and the
method discussed in the preceding paragraph is that we need only a single photocounter for the former.

Photon counting

The voltage pulses supplied by a single photodetector (originating from the generation of a single photoelectron) can be electronically counted in time intervals of
equal length. In this way we obtain the number of photons registered by the detector within the individual time intervals. Of great physical interest are the statistical
variations of the measured photon numbers, which usually differ from case to case.

Finishing this short review about the available technical possibilities for the observation of optical phenomena, let us ask the following question: To what extent
does the photoeffect deliver a watertight proof for the existence of photons as spatially localized energy lumps introduced in the Einstein light quanta hypothesis?

5.3 The photoeffect and the quantum nature of light

The photoeffect is described within quantum mechanics very similarly to the process of light absorption discussed in Section 5.1. The reason is that we can speak
of well defined atomic states even when the electron is completely detached from
the atom. These states form, so far as the energy is concerned, a higher energy
continuum which is attached to the discrete energy spectrum. Its energy is composed additively of two parts, the value E_c marking the lower edge of the continuum and the kinetic energy of the essentially free electron:

$$E = E_c + \frac{1}{2}mv^2, \tag{5.5}$$

where m is the mass and v is the velocity of the electron. The continuous character
of the spectrum – the fact that in this region any value of the energy is allowed
– is caused by the fact that the kinetic energy of the free electron is not subject

to any type of "quantization rule," and it can be safely assumed that the classical description applies such that any non-negative value of energy can be taken.

The quantum mechanical description makes no distinction between transitions between two discrete energy levels or between two states, one of which belongs to the discrete and the other to the continuous part of the spectrum. In particular, Equation (5.1) relating a certain transition to a particular atomic frequency keeps its validity. Using Equations (5.1) and (5.5) we obtain for the resonance frequency of the ionization process the relation

$$\nu = \frac{1}{h} \left(E_c - E_0 + \frac{1}{2}mv^2 \right), \tag{5.6}$$

where E_0 is the energy of the initial atomic state (usually identical to the ground state of the atom). Because the energy difference $E_c - E_0$ is simply the ionization energy, we immediately obtain in this way the Einstein Equation (2.1). The energy balance obtained from the quantum mechanical point of view, certainly does not imply the photon picture, it just reflects the general resonance character of the interaction between radiation and matter. In fact, Equation (5.6) can already be found in the semiclassical theory, i.e. when the atoms are treated quantum mechanically and the radiation is treated *classically*!

What can be said with certainty is that the unrestricted validity of the energy conservation law for individual processes, such as the release of a photoelectron, implies an extraction of energy from the radiation field of magnitude $E_c - E_0 + \frac{1}{2}mv^2$, which equals $h\nu$ according to Equation (5.6).

Can it be that the particle nature of light is only mimicked by the detector? The atom participating in the elementary process can, due to its structure dictated by quantum mechanics, only either take from the radiation field of frequency ν the energy $h\nu$ or not act at all. Is it not possible that the electromagnetic field does not have a grainy structure at all, but is instead similar to a soup which is "quantized" only when it is eaten portion-wise with a spoon? Is classical electrodynamics right after all by ascribing to light a spatially continuous energy distribution? And is it true that an atom, following the rules of classical theory, collects energy from the field until it has gathered enough energy ($h\nu$) required for the transition?

Our experience tells us that the classical concept cannot withstand scrutiny. Let us estimate the "accumulation time" required by the atom to collect the energy $h\nu$ from the radiation field according to classical theory! (We follow an argument given by Planck (1966).)

Let us assume the atoms participating in the photoeffect to be "densely packed," as in a solid. Obviously the atoms placed right at the surface and thus being hit directly by the incident light are affected the most by the radiation field. In the most favorable case they can absorb the whole incident energy, and we will assume that this is indeed the case to estimate the *minimal duration* of the accumulation process. Let us make the additional assumption – motivated by quantum mechanics – that all the atoms are initially (at the moment when the irradiation begins) in the same state and, apart from boundary effects, that none of the surface atoms is in any way preferred. This means that their behavior under the influence of the radiation field (we consider a plane, quasimonochromatic wave, incident orthogonally to the surface of the solid) must be identical. This is the general consequence of the deterministic character of any classical theory.

The surface atoms cannot take energy from each other. The energy at their disposal is (assuming dense packing) given by the flux through their geometrical cross section. Introducing a photon current by counting the energy of the incident field in units of $h\nu$, we find the following: for a given photon flux density j, in total jF photons per second flow through the atomic cross section F; i.e. it takes time

$$T_{\min} = \frac{1}{jF} \tag{5.7}$$

for energy corresponding to that of one photon to flow through the atom. This time is also an estimate for the shortest possible duration of the absorption process in which the atom swallows an energy quantum $h\nu$; i.e. $T_{\min}$ is the minimal value of the classical accumulation time.

To form an impression of what might be the values of a realistic photon density flow, let us consider the most natural source – the sun. Let us assume we cut out, using a filter, green light at a wavelength of 500 nm with a frequency width of 1% of the frequency. The corresponding photon flow density is of the order of $10^{19}/(m^2\,s)$, which is easily calculated from Planck's radiation formula and the known parameters for the sun (surface temperature =6000 K, angular diameter 32 arc minutes). Using these parameters, Equation (5.7) gives a minimum duration of the accumulation process of 10^{-1} s for an atomic cross section of $10^{-18}\,m^2$.

This calculated time is relatively large; however, at considerably smaller intensities we obtain fantastic accumulation times. Actually, the photoeffect sets in at much lower intensities. This is known from direct experience, namely our visual perception, where the elementary process is of photoelectric nature. In fact, the response threshold of the human eye (for green light) corresponds to an energy flow of 5×10^{-17} W (see, for example, Wawilow (1954)), which implies a photon flow of 120 photons per second. A photon flow density of, say, $10^{10}/(m^2\,s)$ is certainly sufficient to initiate the generation of photoelectrons. However, according to Equation (5.7) it would take at least 10^8 s; i.e. we would have to wait longer than three years to see the first photoelectrons being detected – a rather bizarre result.

These arguments show us that our everyday experience of visual perception is a convincing argument in favor of the photon nature of light. Were classical theory correct, we would not be able to perceive optically feeble objects at all – the overwhelmingly beautiful night sky full of stars, which has influenced so strongly the development of physics, would be hidden from our eyes. In addition, the process of seeing would be very strange. The different brightnesses of objects would be perceived through the different times it would take to recognize them. After that, however, each object (if we consider the idealized case that the intensity of incident light from the object is the same everywhere on the retina) would appear with the same brightness because all the receptors affected by the light from the object would respond in the same way and at the same time. The vision process becomes understandable only with the photon picture, which is similar to that of a shower hitting the retina, whereby a sensitive element (a rhodopsin molecule in the rod) remains either completely unaffected or receives the required energy "at one blow."

One of the reasons for the erroneous estimation of the accumulation time given by the classical theory is the assumption of densely packed atoms. In this case, atoms block each other, so preventing the atoms from absorbing the electromagnetic energy from a larger surrounding area. In fact, such a "suction effect" for atoms far from each other (or if, for high densities, only a few atoms participate – for whatever reasons – in the interaction with the radiation field, which amounts to the same thing) is recognized from classical electrodynamics.

We show in the following that classical theory leads under such conditions to a contradiction with our experience; however, in this case the discrepancy between the theoretical prediction and the experimental results is not as dramatic as in the previous case of densely packed atoms. We start again with the estimation of the accumulation time within the classical description.

The simplest model of classical electrodynamics for the elementary process of absorption (independent of whether it is excitation or ionization) is that of an electron elastically bound to an attractive center. Hence we are dealing with a harmonic oscillator that is put into resonant oscillation by the radiation field of frequency ν. We are interested in the time T needed for the oscillating electron to accumulate an energy (given by the sum of the kinetic and the potential energy) that equals $h\nu$, the atom thus having absorbed a whole light quantum. (We will not analyze the question of how the electron is released from the atom in the case of the photoeffect.) A simple calculation made by Lorentz (1910) gives for the accumulation time the value

$$T = \frac{\sqrt{8mh\nu}}{eE_0},$$
(5.8)

where E_0 is the amplitude of the electric field strength $E(t) = E_0 \cos(2\pi\nu t - \varphi)$ at the position of the atom, e is the electronic elementary charge and m is the mass of the electron. The atom theory of Bohr (in which we consider a hydrogen-like atom) allows us to replace the variable $2\pi v m$ by $\hbar/r_0^2$, where r_0 is Bohr's radius. (Because we are interested only in the order of magnitude of T, we can assume for simplicity the principal quantum number to be equal to unity.) Thus, Equation (5.8) can be replaced by

$$T \approx \frac{\sqrt{2}h}{\pi D E_0}.$$ (5.9)

Here we have introduced the parameter $D = e r_0$, which represents, at least in order of magnitude, the atomic dipole moment.

Expressing in Equation (5.9) the amplitude E_0 of the electric field strength by the time averaged photon flux density j,

$$E_0 = \sqrt{2hvj} \; \sqrt[4]{\frac{\mu_0}{\epsilon_0}},$$ (5.10)

(μ_0 is the vacuum permeability), we obtain the final expression

$$T \approx \frac{1}{\pi D} \sqrt{\frac{h}{vj}} \; \sqrt[4]{\frac{\epsilon_0}{\mu_0}}.$$ (5.11)

Comparing with the previously obtained result of Equation (5.7), we see that the dependence on the photon flux density is much weaker. The absolute values for the accumulation time are indeed much smaller than those calculated previously. For $j = 10^{19}/(m^2 \, s)$ (green spectral part) we find for T the rather small value 3×10^{-7} s (the earlier estimate was 10^{-1} s), while for $j = 10^{10}/(m^2 \, s)$ the earlier absurd value of 10^8 s drops to 10^{-1}s. The dramatic decrease in accumulation time can be understood if the atom – when the environment does not hinder it – is drawing the energy from a volume considerably larger than its own dimensions. The mechanism responsible for this effect is well known from antenna theory: the incident wave induces an atomic dipole moment which by itself emits a wave. The wave emitted by the absorber, however, does not transport any energy away from the atom – on the contrary, the wave interferes with the incident wave (there is a fixed phase relation between the two waves which is a characteristic for the absorption process) in such a clever way that even the energy flowing around the atom at larger distances is directed towards the atom. Figure 5.1 shows the shape of the energy flow, time averaged over several light periods (for details see Paul and Fischer (1983). The "suction effect" of the atom is clearly seen, and is surprisingly strong: energy that has already passed by the atom is diverted back to the atom.

We should point out that the described mechanism works only as long as the phase of the incident wave stays constant. Any change (for instance a jump) in the phase destroys the optimal absorption phase relation between the incident wave

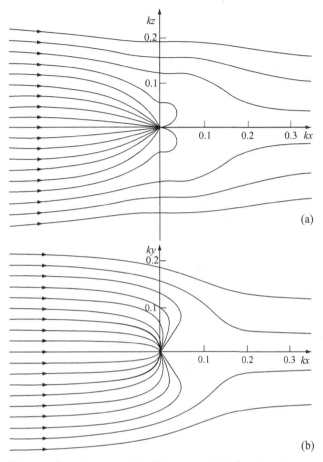

Fig. 5.1. Energy flow into an absorbing atom (a) in the plane spanned by the propagation direction of the monochromatic, linearly polarized plane wave (x) and the oscillation direction of the induced dipole moment (z), and (b) in the x, y plane orthogonal to the former plane (k is the wave number). After Paul and Fischer (1983).

and the dipole oscillation, and the accumulation time becomes much longer. Equation (5.8) is valid only under the assumption that the coherence time of the incident wave is not shorter than the accumulation time.

In fact, the values of the accumulation time predicted by Equations (5.8) and (5.11) are still orders of magnitude larger than the values obtained experimentally. The two fundamental experiments performed in this context are very original, and we do not miss the opportunity to describe them in some detail as they have been (unjustifiably) almost forgotten.

The main experimental challenge is essentially the measurement of the anticipated extremely short times that pass between the beginning of irradiation and the appearance of the first photoelectrons. One might think that sophisticated ultra-short time measurements, which have only just recently been developed, would be necessary. However, a surprisingly simple solution of the problem was found in the 1920s by the American researchers Lawrence and Beams (1927). They generated, with the help of a spark discharge, a light flash which was sent onto the cathode of a photocell. The cathode was equipped with an additional grid. At the moment of the discharge there was a positive voltage between the grid and the cathode so that all the released electrons were drawn away from the cathode. An appropriate circuit arrangement guaranteed that the discharge simultaneously caused a reversal of the grid voltage. The simple fact that an electric pulse is propagating along a wire with the velocity of light was used to delay the voltage change by an interval τ, compared with the beginning of the irradiation of the cathode. This was accomplished simply by using a wire that was longer than the optical path. The problem of time measurement was thus reduced to the measurement of length! The now negative grid voltage blocked the cathode photoelectrons. The registered photoelectrons could have been released at a maximum time delay τ if compared with the beginning of the irradiation. Lawrence and Beans concluded that the photoelectrons were released *simultaneously* with the beginning of the irradiation. They reported a precision of the apparatus with respect to the delay time to be 3×10^{-9} s. Hence, the accumulation time – if there is any – can be 3 ns at maximum.

The aim of the second experiment carried out by Forrester, Gudmundsen and Johnson (1955) was completely different, namely the detection of interference – in the form of beats – between two thermally generated monochromatic light waves with slightly different frequencies, and it is by itself of fundamental importance. (In Section 7.4 we will consider the experiment in detail.) What is of interest for us is the conclusion drawn by the authors, namely that the observed beat signal (in the photocurrent) at a difference frequency of 10^{10} Hz can be formed only when the photoelectrons follow exactly the fluctuations of the light intensity, which means that the accumulation time must be significantly shorter than 10^{-10} s.

Because the classically calculated accumulation time is highly dependent on the intensity of the incident light, we need, for a comparison with the predictions of the classical theory, data about the intensity. Fortunately these were measured by Forrester *et al.* (1955). The energy flux density on the photocathode surface was found to be about $0.8\,W/m^2$, for which (for a wavelength of 546.1 nm) a photon density flux of $2 \times 10^{18}/(m^2\,s)$ follows. The accumulation time (given by Equation (5.11)) for this value is approximately 7×10^{-7} s. The obtained value is too large by at least four orders of magnitude, and the statement about the incompatibility of the classical description with experience can be considered as proven.

Interestingly enough, a formula almost identical to Equation (5.9) can be obtained using a quantum mechanical description of the absorption process for a strong, monochromatic (and hence coherent) field.[1] (The important point is that the atom is treated quantum mechanically.) Actually, one obtains the relation

$$T_{\text{qu}} = \frac{h}{2DE_0},\tag{5.12}$$

for the time when the atom is with 100% probability in its upper state, i.e. when it has absorbed with certainty an energy quantum $h\nu$ from the radiation field starting from the lower state (see, for example, Paul (1969)). The symbol D stands now for the quantum mechanical "transition dipole moment," i.e. the non-diagonal element of the electric dipole operator with respect to the two atomic levels participating in the transition, and we have assumed that the mean lifetimes of the two atomic levels (determined by spontaneous processes and collision processes) are considerable larger than the "transition time." In fact, such an agreement between the classical and quantum mechanical predictions is to be expected from the correspondence principle!

One might question what is the essential difference between the two types of descriptions. The answer is, it is the intrinsically statistical character of quantum mechanical predictions, as already explained in Section 4.1, making a deterministic description of individual microscopic single systems impossible in principle (in contrast to classical theory). The difference is best seen when we consider times t shorter than the accumulation times given by Equations (5.9) or (5.12). The classical as well as the quantum mechanical theory agree on the statement "The atom has absorbed up to time t only a certain fraction of a single quantum $h\nu$," but the meaning of it is completely different for the two theories. Within the classical description, the statement has to be taken literally in the sense that *each single* member of the ensemble of atoms (being all initially in the same state) absorbed no more and no less than the particular energy amount. However, because such an amount is not sufficient to start ionization (consider the photoeffect), no photoelectrons can be detected at this moment. The quantum mechanical statements, on the other hand, refer to the behavior of the atoms observable *on average* in an ensemble; the individual atoms necessarily do not act in the same way, but instead each follows its own destiny. The mean value (say $\frac{1}{4}h\nu$) of absorbed energy actually comes about such that, at the considered time t, 25% of the atoms have absorbed one whole quantum while the remaining 75% come out empty handed.

---

[1] In principle, this is not surprising, as the quantum mechanical equations are just a very clever "translation" of the classical ones, and anyway the classical formula Equation (5.9) was derived with a loan from quantum mechanics (by employing Bohr's model).

The separation of the whole ensemble into two sub-ensembles assumes, strictly speaking, the existence of a suitable measurement procedure which "finds out" the energy states of the atoms. The photoemission itself can be already considered as a substantial element of such a measurement process. In fact, the release of a single electron is an irreversible process because the electron is moving away from the atom, and from it, for example through an avalanche process (see Section 5.2), we can obtain a macroscopic signal.

To recap, let us again state the fundamental difference between the classical and the quantum mechanical descriptions of the photoeffect. In the classical description, the fate of the atoms is predicted in fine detail – from the initial conditions and the external influences it can be uniquely determined, and hence it is identical for all the atoms that have the same initial conditions. In the quantum mechanical description, the life of an atom is a lottery. Similar to the case of radioactive decay, where some "daring" atomic nuclei decay immediately while others take their time, some of the atoms are ionized after a very short time whereas others are ionized later; the ionization events are distributed over the whole time scale. The sudden emission of electrons by atoms due to light irradiation setting in after the elapse of the accumulation time in the classical picture – all the atoms eject an electron more or less simultaneously – does not take place. The weird quantum mechanical rules of Nature guarantee that the process of ionization is "resolved" into a great number of single events taking place at different times. We would like to emphasize once again that this is a general property of the quantum mechanical description of Nature; the "dice playing," which Einstein believed he could not expect of God, actually takes place.

We cannot resist pointing out the amazing consequence of this peculiarity of quantum mechanics which changed so fundamentally our concepts about the micro-processes. We consider as an example a problem which is currently of interest in particle physics, the question of the stability of the proton. Fundamental considerations led to the concept that the proton is not a stable particle but disintegrates on average after about 10^{31} years. This inconceivably long time, exceeding the age of the universe by 21 orders of magnitude, seems to make the question highly academic – a speculation having no relation to physics as a science of experience. In fact, we should realize that if we consider the huge number of protons contained, for instance, in one cubic kilometer of water, some – within days or weeks – might decay, and such a process would be detectable with good measurement equipment and a lot of patience – at least it is within the realm of experimental possibility! The performed experiments indicate, however, that the decay time of the proton, if it is really unstable, must be larger than 10^{31} years. This is, let us emphasize once again, a real experimental result!

Let us return after this short detour to the photoeffect! The difference between the classical and the quantum mechanical description explained in detail in the previous text manifests itself also macroscopically, namely in the time dependence of the photocurrent. The classical theory predicts an alternating current, while quantum mechanics – in excellent agreement with experience – predicts a direct current.

This example shows that we must be very careful with the following frequently drawn conclusion: the correspondence principle applies to ensemble mean values; macroscopic measurements always deal with this type of mean values, and hence, in the macroscopic regime, the quantum mechanical description goes over into the classical. In fact, we *do not* measure the average value of the absorbed energy in the photoeffect (which is as explained the same in both cases). The conclusion applies only to the attenuation of the incident light as a result of the photoeffect.

The arguments presented in this section leave no doubt that one of the aspects of radiation is its particle nature. The electromagnetic energy of a radiation field (of small intensity) cannot be imagined as being more or less evenly distributed in space because in such a case the atom would not be able to "collect" within such a short time the required amount of energy $h\nu$.

Independent of all the theoretical considerations, we can adopt the following pragmatic point of view: there are instruments (photodetectors) which reliably indicate that an energy amount of $h\nu$ was taken from the field. Such an event is called "detection of a photon." Because the process inevitably leads to the destruction of the photon, we do not learn too much about the real "existence" of the photon. So it seems reasonable to turn our attention to the "birth" of a photon. This takes place during the process of spontaneous emission.

6

Spontaneous emission

6.1 Particle properties of radiation

One of the most important properties of macroscopic material systems is their ability to emit radiation spontaneously. According to quantum mechanics, the emission process is realized in the following way: an atom (or a molecule) makes a transition from a higher lying energy level (to which it was brought, for example, by an electron collision) to a lower lying energy level without any noticeable external influence (in the form of an existing electromagnetic field), and the released energy is emitted in the form of electromagnetic radiation. The discrete energy structure of the atom dictated by the laws of quantum mechanics is imprinted also on the emission process (quantization of the emission energy), since the energy conservation law is also valid for single (individual) transitions. Hence, a single photon, in the sense of a well defined energy quantum, is always emitted.

The emitted quanta can be directly detected by a photodetector. (Strictly speaking, identifying a registered photon with an emitted one is possible only when it is guaranteed that the observed volume contains only a single atom. (For details see Sections 6.8 and 8.1.) Under realistic conditions, the experiment can be performed in the following way. First, a beam of ionized atoms is sent through a thin foil; the emerging beam then consists of excited atoms. (This procedure is known as the beam–foil technique.) A detector is placed at a distance d from the foil to detect light emitted sideways by the atomic beam (Fig. 6.1). The setup is used to measure the relationship between the frequency with which the detector responds and the distance d. (The detector is moved along the beam, providing different values for d.) Obviously a single event registered by the detector represents the absorption of an energy quantum at time $t = d/v$ (with v being the velocity of the atoms) counted from the instant of the excitation of the atom. Because the response of

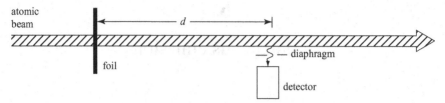

Fig. 6.1.　Beam–foil technique.

the detector indicates that the atom emitted *all* of its excitation energy,[1] we can conclude that the atom at time t (and also later if no additional transitions follow) is *with certainty* in the lower state; i.e. it has completed its transition.

Another experimental method employed in the observation of the quantum-like emission uses the possibilities offered by modern laser technology, which allows the generation of extremely intense and ultra-short light pulses (of the duration of picoseconds down to femtoseconds).[2] Using pulses of appropriate frequency it is possible to excite atoms which have initially been in the ground state at a chosen instant – more precisely, within a time interval much shorter than the mean lifetime of the excited level. With the help of a photodetector we can then determine the time t (that has elapsed since excitation) at which the first photon is detected. By repeating the experiment many times we can determine the response frequency of the detector as a function of time t. (To do so we need a detector with a response time much shorter than the duration of the whole emission process. In addition, the experiment must be performed for small light intensities so the detection of photons that have been emitted later is possible.) In this way we obtain the same information as in the case of the beam–foil technique.

These two experiments present the atomic transition as discontinuous: as an instantaneous and jump-like process. In fact, this picture was introduced previously by Niels Bohr. It seems as if for a certain time interval nothing happens; the atom remains in its initial state until abruptly, at an unpredictable moment t, it "decides" to make the transition. We know from our experience that the response frequency of the photodetector, as a function of time t counted from the earliest possible arrival time t_0 of a light quantum (where $t_0 = t_e + t_1$, t_e being the "instant" of excitation and t_1 being the time needed by light to reach the detector from the emitter), has an exponential form similar to that describing radioactive decay. This means that the probability of the detector registering a photon within

---

[1] Strictly speaking, this conclusion is justified only when the experimental conditions guarantee that the detector "faces" only one atom, during its response time; otherwise the absorbed energy could originate from several atoms.

[2] In many cases we may also work with conventional sources such as mercury high pressure lamps complemented with a chopper, for example a rotating disk with small holes which "chops" the emitted light into a sequence of pulses.

the time interval $t \ldots t + \mathrm{d}t$ is given by

$$\mathrm{d}w = \mathrm{const.} \times \mathrm{e}^{-\Gamma t} \, \mathrm{d}t. \qquad (6.1)$$

On integrating Equation (6.1), we find that, from all the registered events, that fraction for which a photon is detected at a time later than t equals $\exp\{-\Gamma t\}$. From this we can conclude that – assuming a jump-like elementary emission process – a fraction $\exp\{-\Gamma t\}$ of the initially excited atoms is, at time t, still in the upper (excited) state. This means that Γ^{-1} can be interpreted as the mean lifetime of the excited level. In fact, it recently became possible to visualize the quantum jumps of individual atoms. The observation uses resonance fluorescence induced by a (resonant) incident laser wave. During the process a fair number of photons are emitted to the sides. We can imagine the process of scattering in the following way. An atom, which is initially in the ground state, is excited by the intense radiation and reaches a higher energy level. The energy obtained by the excitation is radiated away in a random direction in the form of a photon.[3] In complete analogy to spontaneous emission, in this process the atom takes its time: the frequency distribution for the emission "instants" is exactly the same as in the previous case (see Equation (6.1)). On average, it takes Γ^{-1} seconds until an emission takes place, and it is of particular importance that the mean lifetime of the corresponding atomic level, Γ^{-1}, is a parameter independent of the intensity of the incident radiation. (In contrast, the atom is "pumped" faster from the ground state to one of the excited states as the intensity of the pump wave increases.) After the emission is completed, the atom is again in the ground state and the process can start again. Longer illumination times result in a larger number of emitted fluorescence photons. Let us now consider the case when the upper level can, in addition, decay into a third, lower lying, level through a transition associated with spontaneous emission. From this it can return through additional spontaneous transitions to the ground state.

Let us assume that the first transition is very weak (for example, it might be a quadrupole transition); in such a case, the competition between the transitions will be decided in favor of resonance fluorescence – however, this will not last forever: at an unpredictable moment the weak transition "takes its turn." The consequence is an abrupt interruption of the fluorescence – this lasts until the atom has reached the ground state and thus can be "driven" again by the laser radiation. The result is that we can observe an irregular sequence of bright and dark periods of resonance fluorescence (intermittent fluorescence; see Fig. 6.2), and the interruption of the fluorescence indicates that a spontaneous transition, the famous quantum jump,

---

[3] We employ without hesitation the photon picture because we assume that the scattered photons are detected using photodetectors.

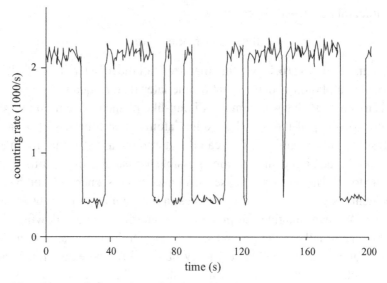

Fig. 6.2.   Observed alternation of dark and bright periods in resonance fluorescence (from Sauter *et al.* (1986)).

has just happened. Because during the bright phase a large number of laser photons are scattered, the fluorescence radiation can be viewed as a *macroscopic* signal, and its "switch off" can be reliably detected. In this way we are able to detect, in an indirect way, the spontaneous emission of a light quantum with *certainty*, whereby we can even use low efficiency detectors. Hence, the described technique is well suited to determine the mean lifetime of an excited level with respect to the weak transition. To do so we have to measure the dependence of the fluorescence interruption rate on the intensity of the fluorescence radiation (Itano *et al.*, 1987).

However, this method for observing quantum jumps is applicable only when we deal with a single atom. (The random character of the bright and the dark phases of the fluorescence radiation leads to an essentially constant total radiation when the contributions from several atoms are superposed.) Fortunately, modern state of the art detectors allow us to fulfil this precondition almost ideally. It is possible to capture an individual ion in an electromagnetic trap, the so-called Paul-trap, and keep it there, in principle, for an infinitely long time (Neuhauser *et al.*, 1980), an essential trick being the electromagnetic "cooling" of the ion. This means that the back and forth oscillations of the ion are damped away by irradiating the ion with laser light of frequency slightly below the corresponding resonance frequency of the ion, thus utilizing the radiation pressure.

In fact, the fluorescence technique can be further improved. To this end, a second laser is used to drive the weak transition $1 \leftrightarrow 2$ which is coupled to the ground state (the so-called V-configuration; see Fig. 6.3). Then the absence of

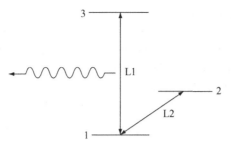

Fig. 6.3. Two coupled atomic transitions induced by laser radiation L1 and L2. Intermittent resonance fluorescence is observed on the strong transition $1 \leftrightarrow 3$.

the resonance fluorescence associated with the strong transition $1 \leftrightarrow 3$ indicates that the ground state has been emptied by an *induced* weak transition $1 \rightarrow 2$ (also jump-like). After a while, the atom returns, either through an induced or spontaneous transition, to the ground state, and the fluorescence is again "switched on."

The described technique might be relevant for the construction of extremely precise frequency standards. With the weak transition we have at our disposal an extremely narrow emission or absorption line. A properly chosen line of this form could be used for the definition of a frequency standard. Its realization by a laser beam could be achieved simply by tuning a suitable laser to the weak transition, whereby resonance would be indicated by the appearance of dark periods in the resonance fluorescence.

Finally, let us mention that, with the help of the described experimental method, quantum mechanical energy measurements may also be carried out. To this end, instead of continuous laser light resonant with the strong transition we use a sequence of very short pulses, each of them so intense that it generates a large number of fluorescence photons. If it is otherwise known that the atom is in level 1 or 2 (of special interest is the case of a superposition of the corresponding two energy states), then each of the pulses "interrogates" the atom to discover which energy level it is in. The appearance of fluorescence indicates the atom to be in level 1, while its absence allows the conclusion that it is in level 2. A recent successful experimental demonstration of the so-called quantum Zeno effect, explained in Section 4.1, was accomplished using this method (Itano *et al.*, 1990).

6.2 The wave aspect

Instead of counting the emitted photons (during the detection process the photons vanish and we cannot say anything more about their physical properties), we can also perform experiments which reveal their wave aspect.

For example, it is possible to perform interference experiments, in principle, for arbitrarily small intensities, as will be explained in detail in Section 7.2. (The exposure time of the photographic plate has to be long enough to obtain an interference pattern.) This implies that a single photon must be able to interfere (with itself). Let us consider, for example, Young's interference experiment (see Section 2.3), in which each of the photons somehow finds out that the diffraction screen has two holes because it acts accordingly, as can be seen from the interference pattern on the observation screen.

A physical explanation of this effect cannot exist without making use of the picture of an extended wave (in the transverse direction). The interference pattern allows us also to draw conclusions about the pulse length corresponding to a photon. Let us look at the region on the observation screen where the interference fringes are at the boundaries of visibility. There, the difference between the two path lengths (counted from the first and second hole, respectively) approximately equals the length of the "elementary wave." (The head of the wave coming from one hole meets the tail of the wave coming from the other hole so that interference is still possible.) A convenient method for measuring the length of the elementary wave is offered by the Michelson interferometer. One has to vary the length of one of the interferometer arms until the visibility becomes poor.

From the experimental point of view, there is no doubt that under certain experimental conditions the photon must be treated as a spatially extended wave. However, such a wave packet needs for its formation a finite time because the duration of the emission process Δt determines, according to the formula $l = c\Delta t$, where c is the velocity of light, the length l of the pulse (in the propagation direction).[4] This fact is, at least in the classical description, obvious. The quantum mechanical description does not invalidate this conclusion because the relation between l and Δt in question is a direct consequence of two principles, valid in the classical theory as well as the quantum mechanical theory: (a) the interaction between an electric dipole (idealized as point-like) and the electromagnetic field is local (i.e., the dipole oscillation and only the simultaneous electric field strength residing at the same position affect one another directly); and (b) the electromagnetic excitation propagates in space with the velocity of light.

The observed wave properties of spontaneously emitted light are understandable only by assuming a *continuous* emission process of a certain finite duration. Information about the duration can be obtained from the aforementioned interferometric measurement of the elementary wavelength or from the frequency spectrum

---

[4] Usually, the wave propagates simultaneously along different directions; in the most frequent case of an electric dipole transition, it does this according to the direction characteristics of the Hertzian dipole radiation (for details see Section 6.5).

of the emitted radiation. The experimental as well as the theoretical analysis shows that the following relation between the natural linewidth and the mean lifetime $T = 1/\Gamma$ of the excited state holds:

$$\Delta \nu \, T = \frac{1}{2\pi}. \tag{6.2}$$

Comparing this result with the fundamental relation Equation (3.22) between the linewidth and the duration of a pulse (which remains untouched by the quantization of the radiation field), we come to the conclusion that the duration of the emitted elementary pulses – and necessarily also of the radiation process itself – is given by the mean lifetime of the upper level.

Equation (6.2) is often seen as a consequence of the general quantum mechanical relation Equation (3.23) between the energy and the time taken to measure it. The presence of an uncertainty in the frequency of the emitted light (according to Einstein's relation between energy and frequency of a light quantum $E = h\nu$, this is equivalent to an energy uncertainty $\Delta E = h\Delta \nu$) is interpreted in such a way that the upper level has the same energy uncertainty and the latter is directly transferred to the emitted radiation field during emission. A more detailed argument is as follows. Because of the finite lifetime T of the upper level we cannot, *in principle*, perform energy measurements (on average) with a duration longer than T. According to the uncertainty relation Equation (3.23) we cannot, as a matter of principle, determine the energy (on an ensemble of atoms to which all quantum mechanical statements are related) with infinite precision, but only up to a possible error of the order $\Delta E = \hbar/T$. As a consequence, we can think of the energy uncertainty as a "real property" of the atomic level. It finds its way into the energy of the photon, which is what is stated by Equation (6.2).

The described approach has two great advantages. First, it does not make explicit use of the assumption that the "decay" of the excited state is caused only by the emission process. Hence, it is also applicable to cases when, besides spontaneous emission, additional mechanisms (for instance inelastic collisions) are also at work which shorten the mean lifetime. Nevertheless, Equation (6.2) should be valid also under these circumstances. It predicts the appearance of a broader linewidth, and, in fact, this effect has long been known as the pressure broadening of spectral lines – caused by collisions, between atoms or molecules in gases, that become more frequent with increasing pressure. (A detailed picture of this process will be given in Section 6.3.) Secondly, the above argument is easily generalized to the case when not only the upper but also the lower level is instable. The atom can leave this level either by a coupling to a lower lying level (associated with spontaneous emission) or by another, additional, possibility. Then the energy uncertainty at the disposal of the photon, $\Delta E = h\Delta \nu$, is given as the *sum* of the uncertainties of the the two levels; i.e., with the two levels labeled 1 and 2, the relation

$$\Delta \nu = \frac{1}{h}(\Delta E_1 + \Delta E_2) = \frac{1}{2\pi}\left(\frac{1}{T_1} + \frac{1}{T_2}\right) \tag{6.3}$$

holds. This result is also in very good agreement with our experience.

Let us mention one additional fact, known from the physics of microwaves, which supports the assumption that spontaneous emission is associated with continuous radiation of an electromagnetic wave. It deals with spontaneous emission of microwaves, but the radiating atom or molecule is not placed in free space (as assumed up to now) but is put into a resonator whose dimensions are adjusted to the wavelength of the radiation. This means that the eigenfrequency of the resonator coincides with the middle frequency of the emitted radiation. The presence of (reflecting) resonator walls influences drastically the radiation process: the mean lifetime of the excited state is shortened – compared with radiation into free space – by several orders of magnitude! Let us consider as an example the microwave emission of an ammonia molecule at wavelength 1.25 cm (it earned respect because the first maser action was achieved using this molecule; see Gordon, Zeiger and Townes (1954)), where we find a reduction of the order of 10^7 s (approximately 115 days) to 0.1 s. This effect can be physically understood in such a way that the front of the wave is reflected when it hits the resonator wall and, when it passes the molecule again, it stimulates the process of emission (accelerates radiation). It is important that the reflected wave arrives at the position of the molecule with an appropriate phase, and this is guaranteed by tuning the resonator to the transition frequency of the molecule. What applies to the wavefront naturally applies also to the field emitted later, and as a result the molecule experiences an increasing stimulation while the resonator is gradually filled with electromagnetic energy (a *standing* wave is built up). However, we are still dealing with spontaneous emission, i.e. a radiation process starting from the vacuum state of the field.

On the other hand, when the phase of the reflected (backwards running) wave is not favorable with respect to the radiating dipole, the emission process is not stimulated but inhibited. Such a situation is present when the emission frequency is considerably different from the neighboring resonance frequencies. In the ideal case of a resonator with perfectly reflecting walls, the emission would be completely suppressed. Under realistic conditions a drastic increase of the mean lifetime of a Rydberg level[5], compared with spontaneous emission in free space, was observed (Hulet, Hilfer and Kleppner, 1985).

---

[5] Rydberg states are highly excited hydrogen-like states of atoms characterized by a very large principal quantum number n. Due to the large excitation the atoms are strongly "bloated" and have a very large (transition-) dipole moment. The consequence is a strong interaction with the electromagnetic radiation, and they also decay spontaneously very fast (in a transition $n \rightarrow n - 1$). Because the Rydberg levels are very close to each other, the transitions are associated with the emission of microwave photons.

In fact, the stimulating as well as the inhibiting influence of the surroundings on the emission process was also observed in the optical frequency domain. The first experiments demonstrated that a reflecting wall placed at a distance d (of the order of few hundred nanometers) from the emitter has a measurable effect on the emission process in the form of a change in emission duration (Drexhage, 1974). The effect is further enhanced when the emitter is placed between two mirrors. The experiment was performed using a metal mirror onto which several monomolecular layers of fatty acid molecules were deposited, followed by the emitters – also in the form of a monomolecular layer. In this way the desired distance d between the emitters and the metal mirror was adjusted. The boundary between the emitter layer and the air acted as a second (partly reflecting) mirror. When the distance d was altered, a significant change in the mean fluorescence lifetime of the excited level of the emitter molecule (a dye molecule was used) was observed; depending on the phase relation, larger as well as smaller values (compared with the normal situation of an uninfluenced emission process) were measured.

Recently microscopic optical resonators have been successfully constructed using flat mirrors and a piezoelectric element which reduces their mutual distance down to a separation of the order of a light wavelength. Using this setup a pronounced dependence of the mean lifetime of the excited level of an emitter (radiating in the optical domain) on the mirror separation, i.e. on the resonator tuning, could be observed (De Martini *et al.*, 1987).

Let us note finally that the wave character of light spontaneously emitted by an atom can be considered as already proven by the fact that a superposition of many such elementary waves represents an electromagnetic wave process which is classically well described. This is demonstrated beyond any doubt by classical optics because light from conventional (thermal) light sources is of this type.

6.3 Paradoxes relating to the emission process

The experimental facts described in Sections 6.1 and 6.2 force us to imagine the emission process – depending on the experimental conditions – either as a jump-like process connected with the release of an energy quantum, i.e. a localized energy bundle, or as continuous radiation of an electromagnetic wave. When we try to describe continuous radiation classically, we face great difficulties. These difficulties are rooted in the fact that classically – due to Equation (3.4), which describes the relation between the field strength and the energy density – the emission of a wave also means simultaneous emission of energy. In particular this implies that the total available energy of the atom (given by the difference in energy of the energy levels between which the transition takes place) is converted into electromagnetic energy only when the radiation process is fully completed; i.e. only after a time which is larger than the mean lifetime of the upper level. This statement, however, is in contradiction with photoelectric measurements, during which we detect much earlier events in which the whole energy $h\nu$ is fed to the detector via the field.

The discrepancy is even more dramatic in nuclear physics. For the emission of γ quanta from atomic nuclei, lifetimes of the order of years are reported. According to the classical picture, after exciting the particular nuclei we would have to wait a number of years until the nuclei emit the corresponding energy of the γ quanta (this would yield wave trains of the length of light years), while, according to quantum mechanics – in agreement with experience – γ quanta are immediately detectable.

The peculiar property of quantum mechanics that energy is "traded" only in finite minimum portions, can be found also in inelastic collisions. When such a collision happens between an excited atom A and a (not excited) different atom B, then the *whole* excitation energy of A, i.e. a whole energy quantum $h\nu$, is converted into kinetic energy of the collision partners. The only other possibility is that *no* excitation energy is exchanged during the collision. Were only part of the excitation energy converted, atom A would apparently not know what to do with the rest. Fractions of $h\nu$ cannot be emitted for fundamental reasons!

This aspect of the collision process has surprising consequences for the effect of pressure broadening of spectral lines, as mentioned in Section 6.2. We argued that such collisions are responsible for the shortening of the mean lifetime of the upper atomic level. However, it turns out that collision processes in which the excited atoms really lose their excitation energy, and thus reach the lower level earlier, do not have any influence on the properties of the emitted total radiation – except on its intensity – because the atoms involved do not radiate at all!

In contrast to this argument, the collision broadening is easily understandable from the classical point of view: an excited atom radiates for a certain time and loses part of its excitation energy; at a certain moment a collision happens, and the remaining available energy is converted into kinetic energy of the collision partner. The emission process is abruptly terminated, and the whole emitted pulse is shorter than "standard," which implies a broadening of the frequency spectrum (as explained already in Section 3.4).

This is not what the atom experiences from the point of view of quantum mechanics. To obtain an agreement with experience, we have to assume that also such collisions happen which are not associated with energy transfer but disturb the process of radiation and hence cause a frequency broadening. Quite a natural conception of the process – in analogy to the classical description – is that, due to the collision, the emission of the electromagnetic wave (considered as a continuous process) is terminated but the wave nevertheless contains a whole quantum of energy.

A possible interpretation of the collision between atom A and a different atom B is that it is a kind of energy measurement on atom A: when atom B (and also atom A) increase their kinetic energy, it means that atom A was found in the excited state and hence had not emitted a photon; otherwise it was found in the ground state, which implies, at least from the energy point of view, that the process of emission was complete at the moment of collision. A more precise quantum mechanical substantiation of this interpretation is given by the Weisskopf–Wigner solution for spontaneous emission (see Section 6.5).

The result is that, on average, the quantum mechanical and the classical description of the effect of inelastic collisions on the emission process agree (see also Lenz (1924)). In the classical picture, fractions of the energy $h\nu$ are emitted, but as a whole the ensemble of atoms emits the same energy as it does according to quantum theory, and the mechanism leading to the broadening of the spectral line – the shortening of the emitted pulses – is the same in both cases. However, the quantum mechanical description causes considerable difficulties when we try to ascribe to the emitted field a reality in the same sense as in classical physics. Because the atom cannot "know" when (or whether at all) it will suffer a collision, it is certain that the emission process will start as in the undisturbed case. This implies that the atom will continuously emit an electromagnetic wave. A sudden collision which robs the atom of its whole excitation energy implies that the radiated pulse is all at once "without" energy, which is, from the classical point of view, an absurdity. One might be tempted to assume that the atom should "reel back again" the field emitted "in good faith." However, excluding supernatural influences, this must happen not faster than with the velocity of light. This might, under unfortunate circumstances (recall the example of the γ quanta emission taking years), be completely absurd because in the mean time almost anything can happen to the atom! It really looks as if the previously emitted field collapses instantaneously when it turns out that the "expected" energy supply will not be available (similar to a research project which dies instantly when the promised funding is at once cut). As in the case of the accumulation time, we are forced to question the validity of the classical relation Equation (3.4) between field and energy density. It seems that, according to quantum theory, fields can exist without energy and then instantaneously collapse when we "notice" the absence of energy.

6.4 Complementarity

The observations described in Sections 6.1 and 6.2 clearly indicate a wave–particle dualism for the process of spontaneous emission of radiation. It is of importance, depending on the concrete experimental conditions, that only one or the other of the complementary sides of light are revealed at any time. In the following we would like to present a detailed explanation of the *fundamental* impossibility of measuring simultaneously the precise moment of emission and the frequency of the photon.

In the case of the beam–foil technique the situation is immediately clear. To guarantee that the detector functions correctly, it is necessary to ensure, by appropriate optical imaging, that it is reached only by light emitted from a certain part of the beam. The atoms flying by are coming into the field of view of the detector only for a short time. When we consider the excited atoms as oscillators radiating continuously for a certain time (in the form of a Hertzian dipole), then only a small part of the wave reaches the detector. When we replace the detector by a spectrometer, we find an unrealistically large linewidth, which (corresponding to Equation (3.23)) is caused by "cutting" a small piece from the whole wave, and so does not have anything in common with the original frequency spectrum.

In the case of the second method described in Section 6.1, in which the time instant when the detector responds was measured directly, we notice that a frequency measurement leaves the "arrival time" of a photon undetermined, and this uncertainty increases with increasing resolution of the spectrometer. Let us clarify the situation by analyzing a realistic frequency measurement arrangement, consisting of a spectral apparatus (for example the Fabry–Perot etalon described in Section 3.4) with photographic registration of the exiting light. (Modern methods use an array of photodiodes for detection.) Light of different frequencies is typically imaged onto different positions (usually having the form of rings or stripes). However, a single photon can, at best, cause one single blackening spot or the response of a *single* localized detector, respectively. This makes it quite clear that we cannot make any statements about the frequency spectrum of a single photon. We have to repeat the measurement on many photons and obtain from the detected photons – dependent on the position where the photons have been detected – a frequency spectrum, which must be understood as a characteristic of the respective ensemble. Of physical importance in this connection is the fact that we cannot draw any conclusions about the instant when the photon entered the apparatus from the moment at which the photon was detected. In Section 3.4, we discussed at length, using the example of a Fabry–Perot etalon, that the filter effect is primarily due to the multiple reflections between the two silver layers. It causes a stretching of the incident pulse, and, because the photon can be detected at any point with non-zero light intensity, the detection time is spread over the interval corresponding to the length of the *outgoing* pulse.

Basically, from the experimental point of view, we can accept the dualistic picture of spontaneous emission by realizing that it is impossible, in principle, to observe a microsystem "as it is." In fact, as P. Jordan would say, we can see only the traces it has left in the macro-world, and these traces, we must accept, point in one case towards a particle and in the other towards a wave character of radiation.

In the following section we explain how the theoretical description of quantum mechanics succeeds in putting the complementary aspects "under one roof."

6.5 Quantum mechanical description

A satisfactory treatment of (collision free) spontaneous emission within quantum mechanics was presented by Weisskopf and Wigner (1930 a, b) (see also Section 15.3). These two researchers found an approximate solution which, in contrast to results obtained using perturbation methods, is not limited in its validity to short times. In the following we will concentrate on the physical implications of the Weisskopf–Wigner theory.

Quantum mechanically we describe the spontaneous emission process as a transition from the initial state, characterized by an atom in the excited state and no photon present, to the final state, i.e. the situation where everything is already "settled;" this refers to the situation that exists after the photon has been emitted and the atom has lost its excitation energy (completely). We write this as

$$\Psi_2 \Phi_{\text{vacuum}} \rightarrow \Psi_1 \Phi_{\text{photon}}, \qquad (6.4)$$

where Ψ_1 and Ψ_2 are the wave functions corresponding to the upper and lower levels of the atom, respectively, Φ_{vacuum} describes the vacuum state of the electromagnetic field (cf. Section 4.2) and Φ_{photon} describes a state with exactly one photon in free space. An important achievement of the Weisskopf–Wigner theory is that it is able to give an explicit expression for the wave function Φ_{photon}. With it we have, at our disposal, the maximum information about the radiation field that is possible according to the laws of quantum mechanics.

What can be learnt from the wave function about the physical properties of the emitted photon? It can be expressed as a superposition of energy eigenstates of the whole field characterized by the property that exactly one photon is present in a certain mode of the radiation field. The modes correspond, as explained in Section 4.2, to plane waves with a well defined propagation direction, frequency and polarization. Such a result is, in fact, not surprising because basically all the modes of the radiation field couple to the atom. In this way, quantum theory "unites" two contradicting (from the classical point of view) properties. On the one hand, the emission does not take place in a fixed direction; the propagation direction – corresponding to that of a classical dipole wave – is undetermined, making interference experiments possible. On the other hand, we find, when surrounding the atom at a greater distance with (ideal) detectors, that only one detector responds, thus indicating that it has absorbed the whole excitation energy of the atom. In such an experiment, the photon looks like a "light particle" which is ejected by the atom in a definite (random) direction.

From the weight with which the individual modes participate in the superposition Φ_{photon}, we can immediately read out the direction dependence, the polarization properties and the frequency spectrum of the radiation (Section 15.3). It turns out that the direction characteristics are identical to that of classical dipole radiation[6] (see Equation (3.15)), and the same applies also to the polarization properties. Next, the radiation has a Lorentz-like line profile,

$$w(v) = \frac{\text{const.}}{(v - v_{\text{r}})^2 + (1/4\pi T)^2},$$ (6.5)

with

$$v_{\text{r}} = \frac{1}{h}(E_2 - E_1)$$ (6.6)

being the resonance frequency and E_1 and E_2 being the lower and upper level energies, respectively. From the frequency distribution, Equation (6.5), follows the value $(2\pi T)^{-1}$ for the linewidth Δv; i.e. Equation (6.2), which was anticipated in Section 6.2.

The presence of a finite linewidth Δv implies a finite uncertainty of the emitted energy $\Delta E = h\Delta v$.

The wave function Φ_{photon} allows us to gain knowledge about the spatial extension of the emitted pulse and its propagation in space. The corresponding information is obtained by calculating quantum mechanical expectation values of field variables (operators) for this wave function.

Let us start with the simplest, the expectation value $\langle \hat{\mathbf{E}} \rangle$ of the electric field strength operator. The result is zero, which may come as a surprise. However, the result *does not* imply, as one might at first conclude, that the electric field strength itself vanishes; i.e. that an appropriate measurement (which is only hypothetical anyway because no feasible procedure is known) gives in each single case a zero result. If this were so, the measurement of $\hat{\mathbf{E}}^2$ would also give the value zero, which contradicts the fact (discussed below) that the expectation value of $\hat{\mathbf{E}}^2$ does

---

[6] We have to keep in mind that a two-level system has a spatial orientation. To make a dipole transition possible, the (total) angular momenta of the two levels must differ by one (in units of $\hbar$), which means that at least one of them is a sub-level of a degenerate level. To separate the sub-levels from each other so that one of them can be chosen, we need a static, homogeneous magnetic field inducing a Zeeman splitting of the sub-levels, and its direction simultaneously defines the dipole axis. More precisely speaking, this applies to a transition for which the magnetic quantum number (the angular momentum component with respect to the magnetic field direction) does not change. However, in the case of a change by $+1$ or -1, the motion of the radiating electron corresponds to that of a ring current. Also, in this case there is, concerning the direction characteristics as well as the polarization properties of the radiation, a complete agreement between the quantum mechanical and the classical descriptions.

not vanish. Because the quantum mechanical expectation value should be understood as a mean over many individual measurements always performed under the same physical conditions, the statement $\langle \hat{\mathbf{E}} \rangle = 0$ can be understood only in such a way that the phase of the electric field strength is completely random; i.e. a corresponding measurement would yield any result between 0 and 2π with equal frequency. Since we are dealing with a *pure* state represented by a wave function, the uncertainty, as discussed in Section 4.1, is of purely quantum mechanical nature; i.e. it cannot be understood in the sense of classical statistics that "in reality" each single photon has a certain phase that we just do not know.

The expectation value of $\hat{\mathbf{E}}^2$ is, over the whole space, larger than zero but is not the same everywhere. In a certain space-time region G, its value is above the noise level caused by vacuum fluctuations (see Section 4.3). This simply means that the (mean) intensity $\langle \hat{\mathbf{E}}^{(-)}\hat{\mathbf{E}}^{(+)} \rangle$, which is free of the vacuum influence (see Section 5.2), is non-zero (positive) in G. Because the intensity, according to Equation (5.4), determines the response of the photodetector, we can expect that at one of the space-time "points" belonging to region G a photon will be detected by chance when a detector is placed at such a position. The space-time region indicates the space-time structure of the emitted pulse.

Fortunately, the dependence of the quantity $\langle \hat{\mathbf{E}}^{(-)}(\mathbf{r}, t)\hat{\mathbf{E}}^{(+)}(\mathbf{r}, t) \rangle$ on space and time corresponds exactly to the predictions of classical electrodynamics for the space-time dependence of the intensity of a wave train from a Hertzian dipole (once "kicked" and then left to itself). Observing the radiation along a certain direction (as it is perceived by an observer at a certain distance from the atom), we find the following picture. The intensity is that of a shock wave with a vertical wavefront and an exponentially decaying tail, propagating away from the atom; when traced back, it can be seen that the shock wave began at the instant when the system was in the abovementioned initial state. Fig. 6.4 shows an "instantaneous picture" of this process. The intensity falls off to the e-th part at a distance $\Delta R = cT$ from the wavefront, where c is the velocity of light and T is the mean lifetime of the upper level. From the intensity dependence, the duration of the pulse is found to be T, which is exactly what we expect according to the classical description.

The result shows that the Weisskopf–Wigner solution accurately reflects the wave aspect of the emitted light. However, for the real measurement of the space-time structure of the wave – as well as of the frequency spectrum – photodetectors are the only apparatus available which make the photon vanish as a whole.

This implies that the spatial distribution of the emitted elementary wave can be determined only from an ensemble – each single measurement produces only one space-time "point" at which the photon was found, and it takes many such measurements to form gradually the space-time picture of the wave process. In fact, all the real observations demonstrate the particle nature of light. Paradoxically, we

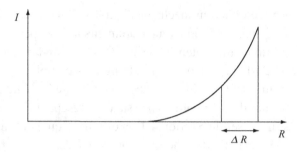

Fig. 6.4. The intensity I of a spontaneously emitted wave train as a function of the distance R from the emitter at a fixed time.

can make statements about the wave properties of photons by performing experiments involving their detection as localized particles; this is the only possibility at our disposal.

The disagreement with the classical description is, to recap, that classically it is impossible to measure the whole energy contained in a pulse at one position because it is, roughly speaking, distributed over a continuously expanding sphere.

The Weisskopf–Wigner theory offers in addition a detailed description of the transition, Equation (6.4), itself by enabling us to write the wave function of the whole system formed by the atom and the field as a function of time. That it is a wave function, and hence a pure state, is a quite general property of the quantum mechanical formalism. We start from a wave function (in our case from $\Psi_2 \Phi_{\text{vacuum}}$); it changes according to the Schrödinger equation, but still remains a wave function. Only through interaction with a macroscopic system acting as a kind of measurement apparatus can the pure state be converted into a statistical mixture. We can conclude that the exclusion of external perturbations enables the whole system to remain in a pure state.

The discussed wave function has the following form:

$$\Psi(t) = e^{-\frac{\Gamma}{2}t}\Psi_2\Phi_{\text{vacuum}} + \sqrt{(1 - e^{-\Gamma t})}\Psi_1\Phi_{\text{photon}}(t) \qquad (6.7)$$

(see Section 15.3), where the wave function $\Phi_{\text{photon}}(t)$ describes, as before, a whole photon (the corresponding energy equals the total amount of the atomic excitation energy). It is this wave function that undergoes a time variation (indicated by the argument t) beyond the "undisturbed" time evolution characteristic of free fields. As is to be expected, this variation is of importance only for those times when the atom and the radiation field mutually interact ($t \leq \Gamma^{-1}$). The properties of the photon described by the wave function $\Phi_{\text{photon}}(t)$ change during the emission process; the particular pulse differs, in agreement with the classical concept, from the one appearing at $t \gg \Gamma^{-1}$ by not being completely "hatched" from the atom (a part of the tail "sticks inside.") Nevertheless, it contains the whole energy $h\nu$ of a photon. Such a "crippled" photon can appear only when the emission process is disturbed from the outside. Such a perturbation is the energy measurement on the atom. According to the laws of quantum mechanics, the so-called "collapse" of the wave function results; i.e. depending on the result of the measurement, the wave function in Equation (6.7) reduces to one of the two terms in the sum (which then has to be normalized again). From a physical point of view, this means that, with certainty, no photon is present when the atom was found in the excited state – this is simply a consequence of the validity of the energy conservation law for single processes; on the other hand, a "crippled" photon was emitted when the atom was found in the lower level. This picture of a perturbed emission process was already proposed in Section 6.3 to explain the collision broadening of spectral lines. There, we interpreted inelastic collisions as a mechanism for the energy measurement of the atom.

At this point we encounter for the first time the important quantum mechanical feature that the measurement on part of a system gives us detailed information about the state of another part of the system, under the condition that the whole system previously existed in an entangled quantum mechanical state, as, for example, represented by the superposition in Equation (6.7). We will discuss this interesting quantum mechanical feature in more detail in Section 11.1.

The most important physical information about the time dependence of the radiation process is contained in the time dependence of the expansion coefficients in Equation (6.7), which obviously expresses the exponential "decay law." However, we have to be cautious with this formulation. According to the laws of quantum mechanics, we should say: when at time t a measurement is performed which "checks" whether the atom is in the upper or the lower state, then we find it with the (relative) frequency $\exp\{-\Gamma t\}$ in the upper state and with the frequency $1 - \exp\{-\Gamma t\}$ in the lower state. Such a measurement process can be, as mentioned several times already, an inelastic collision converting the whole excitation energy into kinetic energy of the collision partners. From such an event we conclude that just before the collision the atom still "possessed" its energy and was "found" by the measurement in the upper state. On the other hand, in order to be able to bring the result into correspondence with the observed collision broadening of the spectral lines, we feel obliged to interpret the absence of an energy transfer in the inelastic collision as an indication that the atom is in the lower state.

Modern laser technology allows the generation of very intense ultrashort laser pulses and thus presents a new opportunity to "find out" the actual level of the atom. In particular cases we can apply a picosecond light pulse of such a frequency that the atom, when in the upper state, will be ionized. The (momentary) occupation of the lower state can be discovered by "pumping" the atom, using a similar pulse, to a short-lived level, from where its spontaneous decay – revealed by detecting the emitted light quantum – indicates a positive result of the measurement.

The role of spontaneous emission can also be played by resonance fluorescence, as explained in Section 6.1. Knowing, in addition, that the atom can be found in only two well defined states, it is sufficient to check the occupation of one of the levels. When the measurement fails to yield a signal, we conclude that the atom is in the other level.

The response of a detector – under the condition that the detector can be reached by the radiation from only one single atom – is a kind of "self-acting" measurement process: we cannot choose the instant of the measurement; we have to leave it to the atom to "decide" when it gives away its energy $h\nu$ to the detector. The important point is that we enable the atom to dispose of all its excitation energy in "one shot." Let us emphasize at this point that, under normal circumstances, the environment plays the role of a measurement apparatus. Each single absorption which is followed by the dissipation of the deposited energy represents a measurement process, even when no "reading out" takes place.

Consider an atom completely surrounded by a detector in the form of a spherical shell having 100% detection efficiency: we would conclude from the absence of a response that the atom – at the earlier (retarded) time instant at which a wavefront would have been emitted to reach the detector surface at the observation time – is still in the upper level. This measurement setup keeps the atom under constant surveillance. It has to report continuously – or, more precisely, every Δt seconds where Δt is the time resolution of the detector – what the energy state of the atom is. Even though the formal description is, in this case, considerably different from that which is valid for the undisturbed radiation (leading to the Weisskopf–Wigner solution) – we have to perform a reduction of the wave function every Δt seconds – the result is again the same exponential decay. Of course, now we cannot make any statements about the emitted photon. It is clear that, using the described measurement apparatus, its properties cannot be observed.

Let us now interpret the wave function in Equation (6.7) according to the known rules of quantum mechanics. Because we are dealing with a superposition state, we come to the following paradoxical statement: the atom *is neither* (for $t \leq \Gamma^{-1}$) in the upper *nor* in the lower level, and *neither* is the photon present *nor* not present. The two situations represented by the two terms in Equation (6.7) are *simultaneously* existing possibilities, but neither of them is "factual." This "fuzziness" of the description provides quantum mechanics with the opportunity to "evade problems" when the classical concept of reality leads to a dilemma, as discussed in the case of collision broadening of spectral lines in Section 6.3. When the atom gives all its excitation energy away during a collision and nothing is left to the radiation field, the atom need not "reel back" any field because, until that moment, no *real* emission took place. The latter emission happened only *virtually* – whatever that might mean!

It is interesting that the virtually present field gradually (as time elapses) becomes real. According to Equation (6.7), the system for $t \gg \Gamma^{-1}$ goes over by itself, just by leaving it undisturbed, to the final state $\Psi_2 \Phi_{\text{photon}}$. We face one of the rare cases in which the Schrödinger equation describes an irreversible process without including a measurement process. (The photon leaves the atom forever.) The formal reason for this behavior is that the atom and the radiation field form a system with infinitely many degrees of freedom (represented by the frequencies, propagation and polarization directions). The situation is completely different when the radiating atom is placed into a resonator with dimensions of the order of the wavelength of the emitted radiation, as discussed in the example of microwave emission in Section 6.2. In this case only a few resonator eigenmodes (in the ideal case only one) are available for the interaction with the atom, and the emission process does not have an irreversible character because the emitted wave is reflected by the resonator walls, so it again reaches the atom, making possible a reabsorption of radiation; this can be followed by another emission, and so on.

Finally, we ask: what is the quantum mechanical picture of a realistic photon spontaneously emitted by an excited atom or molecule? We come to the following conclusion: the photon is a spatially extended object, similar to a classical dipole wave, which is extended in each propagation direction over a distance of the order of c / Γ. It is this spatial extension which makes the known interference effects understandable. However, the situation must not be interpreted in such a way that the energy of the photon is distributed over the aforementioned spatial region because a measurement (using a photodetector) always finds the whole energy at a particular point, i.e. in a localized form. The spatial extension of a photon is formally described by the mean intensity $\langle \hat{\mathbf{E}}^{(-)}(\mathbf{r}, t) \hat{\mathbf{E}}^{(+)}(\mathbf{r}, t) \rangle$, and the detector response probability is proportional to it. Let us repeat again, the photon appears as a localized particle – in the sense of Einstein's photon concept – only in the act of detection, and the naive idea that it is already in this localized state before detection is in open contradiction to experiment.

6.6 Quantum beats

Let us discuss a peculiarity of spontaneous emission which enriches our picture of the photon. This peculiarity is observable when, in contrast to what has been assumed up to now, the "pump mechanism" excites *simultaneously* two (or more) closely lying atomic levels. More precisely, the initial excitation state of the atom is given by a superposition of wave functions corresponding to the different energy

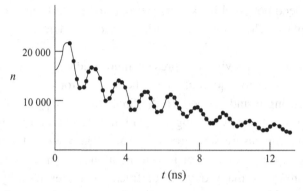

Fig. 6.5. Experimental evidence of quantum beats (n = number of registered photons; t = time passed since the excitation). After Alguard and Drake (1973).

levels. This superposition – a pure state – is related to all atoms present, and we speak of a coherent excitation of the ensemble of atoms.

Experimentally such a specifically quantum mechanical state is achieved either with the aid of the beam–foil technique described in Section 6.1 or by employing short but intense laser pulses. The simultaneous excitation of two different atomic energy levels has the consequence that the two possible transitions into the *same* lower level take place simultaneously. This leads to the fact that the emitted photon – it is, as before, only a single one – has imprinted on it the structure of the atomic excitation: its frequency spectrum consists of two separate lines of finite width. We have to revise our previous concept that the photon has only a *single* frequency of a certain uncertainty. Obviously there are photons "oscillating" simultaneously with two – or even more – frequencies.

Of experimental importance is the effect of this coherent excitation of the atom on the "decay" of the initial atomic state. When, as described in Section 6.1, we measure the number of photons registered by a detector as a function of the time that has elapsed since the moment of excitation, we find – in agreement with theory – strange quantum beats: a sinusoidal oscillation with a frequency given by the distance between the two simultaneously excited levels (in units of h) is superimposed on the exponential decay. In this way, we have a practical method for measuring very small level splittings at our disposal. Fig. 6.5 shows as an example a measurement result obtained using the beam–foil technique.

We should mention that the described beat effect may be understood classically when we assume that two pulses of different frequencies are emitted simultaneously. The intensity of the total radiation, as the result of interference, shows a modulation at the difference frequency (see Equation (3.14)), and this can be detected using photodetectors. The essential point is that we always deal with a single photon consisting of two mutually interfering parts.

6.7 Parametric fluorescence

When discussing spontaneous emission we have, up until now, always considered the radiation of atoms or molecules. Recently, another, completely different, spontaneous emission process was observed. It belongs to the field of non-linear optics, an area that has evolved only because of the development of powerful lasers, and is known as parametric fluorescence or spontaneous parametric down-conversion. It takes place when an intense (monochromatic) wave passes through an appropriately chosen non-linear medium (a crystal).

To explain this process, we require a few additional facts. The non-linearity – or, more precisely, the non-linear dependence of the polarization of the medium on the electric field strength – is of such a form that it allows interaction between three waves with different frequencies; it is this that makes the following process possible. (We will use classical electrodynamics extended by phenomenologically introduced non-linear polarization terms.) By appropriate adjustment of the phases of the three (monochromatic and approximately plane) waves, we can ensure that the wave with the largest frequency, the so-called pump wave, is attenuated during its passage through the crystal, while the other two, usually called the signal wave and the idler wave, are amplified. It is important that the process satisfies the following two conditions (see, for example, Paul (1973a)):

$$\nu_p = \nu_s + \nu_i, \tag{6.8}$$

$$\mathbf{k}_p = \mathbf{k}_s + \mathbf{k}_i. \tag{6.9}$$

The subscripts p, s and i refer to the pump, signal and idler waves, respectively.

Equations (6.8) and (6.9) will become physically understandable when we take a closer look at the physical mechanism responsible for the interaction between the three waves. This is characterized by the property that two waves always induce, on the atoms of the crystal, electric dipole moments oscillating with the sum or the difference frequency of the two waves. In this way a macroscopic polarization is formed in the medium (defined as the sum of dipole moments distributed over a unit volume). In particular, the cooperation between the signal wave and the idler wave gives rise to a polarization oscillating with the sum frequency $\nu_s + \nu_i$, and the pump wave can do work A on it provided its frequency equals that of the polarization, which is, in fact, guaranteed by Equation (6.8). The sign of A depends on the phase of the pump wave with respect to the polarization. Let us assume in the following that conditions are set such that A is positive and, in addition, takes its maximal value.

Equation (6.9) can be explained in such a way that the abovementioned polarization has the character of a plane wave with the wave vector $\mathbf{k}_s + \mathbf{k}_i$. It implies that the pump wave and the polarization wave propagate with the same phase velocity and in the same direction, so that the phase relation between the two waves controlling the energy conversion is preserved along the whole length of the crystal. Hence, Equation (6.9) is called the phase matching condition. When it is not fulfilled, an area in which the pump wave is depleted is followed by an area with the process running backwards; i.e. the pump wave is amplified again, then another area follows in which it is again depleted, etc. and the overall result is that the interaction between the waves is very small.

Equation (6.9) is, however, too strict. In fact, it is sufficient to require that the relative phase between the pump wave and the polarization wave is approximately constant within the crystal volume. Assuming that the pump wave propagates along the z axis and that the length of the crystal is L, we can write the weakened condition for the z component of the wave number vectors in the form

$$\left| k_{\mathrm{p}}^{(z)} - k_{\mathrm{s}}^{(z)} - k_{\mathrm{i}}^{(z)} \right| \leq \frac{\pi}{2L}. \tag{6.10}$$

There are basically two different possibilities satisfy the phase matching condition, namely exploiting parametric interaction of type I or type II. Type I refers to the situation when the signal wave and the idler wave are both the ordinary (or the extraordinary) wave in the crystal. Interaction of type II is present when the signal wave is the ordinary wave and the idler wave is the extraordinary wave, or vice versa. The two waves are then mutually orthogonally linearly polarized.

The process considered up to now – the parametric interaction between pump, signal and idler waves – obviously does not have too much to do with photons. However, the situation changes when the experiment is performed in such a way that we illuminate the crystal with the pump wave but not with the signal wave and the idler wave. According to classical theory, nothing should happen: because two waves are always needed to induce non-linear polarization, at least one additional wave must be present (for example the signal wave, even though its intensity can be arbitrarily small). Such a "nucleus" would then be amplified extremely quickly; "driven" by the polarization generated by the cooperation between the pump wave and the signal wave, an idler wave would build up simultaneously.

The process is, in fact, a spontaneous one; i.e. only the presence of the pump wave is required. During the process, all the pairs of signal and idler waves satisfying Equations (6.8) and (6.9) are excited (for details see, for example, Paul (1973b)). This phenomenon, called parametric fluorescence, is hence of quantum mechanical nature. We can view the corresponding elementary process as the "decay" of a pump photon into both a signal photon and an idler photon. Equation (6.8) multiplied by h then expresses simply the energy conservation. From the point of view of the photon picture, it is important that Equation (6.9) or Equation (6.10) holds for individual spontaneous processes. This is, however, understandable only in the wave picture – in particular, the photon takes notice of the crystal dimensions! – and we cannot help also relating the photon to a wave process in this case. Because the quantity $\hbar\mathbf{k}$ can, on the other hand, be interpreted as the momentum of the photon (see Section 6.9), we can view Equation (6.9) (after multiplication by $\hbar$) as the momentum conservation law.

Of particular physical interest is the fact that during each *elementary* process two photons, in general with different frequencies and propagation directions, are *simultaneously* generated. Hence, when the signal wave and the corresponding idler wave are incident on separate detectors and photons are registered, we find pronounced correlations between the obtained measurement sequences. Using ideal detectors we would, in fact, always find coincidences. The temporal extension of the photons (observed at a fixed point) is – as always in optics – determined by the available linewidth (see Equation (3.22)), which is usually rather large for parametric fluorescence. An experiment which is able to determine the length of the signal or idler photons, respectively, will be described in Section 7.6.

Apart from the aforementioned space-time correlations, there are also correlations in the frequency domain. The energy conservation relation, Equation (6.8), holds, and, when we assume the pump frequency to be very precise (which is indeed the case for a laser), it implies that the measured values of ν_s and ν_i (which fluctuate strongly) are uniquely determined in their sum, i.e. strongly correlated. Finally, using the parametric interaction of type II, we can generate photon pairs with polarization states correlated in a very specific quantum mechanical way. These pairs are well suited for the realization of Einstein–Podolsky–Rosen experiments (Chapter 11).

The photon pairs generated by parametric fluorescence are an instructive example of a system prepared in an "entangled" quantum mechanical state. By performing measurements on one part of the system, we can, it seems, manipulate the other part of the system, a fact that is the essence of the famous paradox of Einstein, Podolsky and Rosen. (For details, see Section 11.2.) In fact, measuring, with the help of a photodetector, the position of the idler photon, for instance, we can predict with certainty the position of the signal photon; however, by measuring the idler frequency with a spectrometer we know also that the signal frequency has a precise value (known when the pump frequency is known). When we impose a limitation on the bandwith of, say, the idler wave – a frequency filter can be placed in front of the detector for this purpose – then this limitation is also transferred to the signal wave.

What has been said above might create the impression that the properties of the signal photon can be changed at will. Because the underlying quantum mechanical "mechanism" is the well known "collapse" of the wave function, which acts "instantaneously", the described effect on the signal photon would take place with superluminal velocity and hence contradict strongly the principle of causality. (We should keep in mind that the signal and the idler photons can, at the moment of measurement, be arbitrarily far apart!) The discussed "action" is hence impossible. What actually occurs in the experiment is something else: we select, through the measurement, from the original ensemble of individual systems prepared in the same way, a sub-ensemble which is physically different from the original ensemble. For example, we measure the idler frequency and only perform experiments with those signal photons for which the measurement on the idler photons has yielded a prescribed value.

It seems interesting to us to point out that the appearance of parametric fluorescence might be viewed as an indication of the existence of vacuum fluctuations of the electric field strength (mentioned in Section 4.3), and, from our point of view, this is a more convincing argument than the often made reference to spontaneous emission of excited atoms or molecules. The fluctuations are well suited to supply the "nuclei" which are necessary in the classical description to initiate the process of parametric interaction.

6.8 Photons in "pure culture"

A light beam, emitted, for example from a conventional light source, can be viewed as a stream of photons. On the one hand, we know that the radiation originates from individual, mutually independent, elementary processes, during each of which a single photon, an energy packet of magnitude $h\nu$, is emitted. On the other hand, the response of the detector tells us that an energy quantum $h\nu$ was removed from the radiation field. Within a naive photon picture, we would assume that the registered photon is identical to that previously emitted by an atom. However, such a picture is incorrect. As will be discussed in detail in Section 8.2, the events indicated by two detectors (for thermal light) are correlated in space and time, which can be understood only if the elementary waves emitted by individual atoms superpose with statistical phases, and photons are found preferentially at the maxima of the intensity of the total field. Hence, the photons do not have an individuality which would – not even in principle – allow us to track their individual "courses of life;" rather, they become completely indistinguishable in the collective.

We arrive at a similar conclusion when we analyze the frequency spectrum of radiation. Usually it is very broad when compared with the natural linewidth characteristic of the elementary process of emission. The effect is called inhomogeneous atomic line broadening. It results from the fact that the individual atoms emit at different center frequencies caused, for example, by the Doppler effect, according to which the motion of the emitter induces a frequency shift. However, it is fundamentally impossible, from the measurement of the total radiation, to gather any information about the spectral properties of the emitted individual photons, in particular the natural linewidth. (This is already so in the classical description: the length of the elementary wave trains constituting the radiation field does not appear in the description of the total radiation in the case of an inhomogeneously broadened atomic line.)

It is possible to speak of an individual photon only when a single atom is its "generator." This means that in experiments we have to take care that the observation volume is filled with just *one* excited atom. When, after a certain time, a photon is observed, it must have come from the atom – we are seeing, so to speak, a photon in "pure culture."

As already discussed in Section 6.1, such a situation is almost ideally achievable with current trap techniques. We can also come close to the desired ideal case when we work with atomic beams, but we have to use a lens to collect radiation from only a small area of the beam and then send it to the detector. The feasibility of this procedure was demonstrated by American scientists in an experiment in which they observed "photon antibunching" (Kimble, Dagenais and Mandel, 1977; Dagenais and Mandel, 1978; see Section 8.4). By choosing the density of atoms and the beam velocity appropriately, the *time averaged* number of atoms in the observation volume dropped below one. However, the method does not exclude that at certain times the observed volume is occupied by two or even more atoms.

Individual photons can also be prepared (approximately) by sufficiently attenuating light from a conventional source or laser radiation (using an absorber or a weakly reflecting or weakly transmitting mirror). In such a situation the probability of detecting a photon within a given (short) time interval is small, but the probability of finding two photons is, in comparison, negligibly small, though not zero.

The experimental situation can be further improved by using generating processes in which two photons are emitted more or less simultaneously. The best candidate for such a process is parametric fluorescence because – in contrast to the successive emission of two photons in an atomic cascade process (see Section 11.1) – the propagation directions of the two photons are strongly correlated, as explained in Section 6.7. When we detect one of the photons in a certain direction, we can be sure that its "partner" set off at the same time and along a known direction. The result is that the motion of the other photon is exactly predictable, as if it were a classical particle. In this way we "generate" a *localized* photon which is available for further experiments.

The information we have about the second photon can be used, for example, to increase considerably the detection sensitivity of an absorption measurement: from the fact that a photon which is expected to arrive at a detector positioned behind an absorber at a certain time does, in fact, not arrive, we must conclude that it was absorbed. An electronic gate is usually used in experiments (Hong and Mandel, 1986). The detector signal indicating the registration of the first photon is used to gate the second detector, which is designed to observe the other photon.

As mentioned in Section 6.7, parametric fluorescence can also be used for the "generation" of single photons with a prescribed frequency.

6.9 Properties of photons

We explained in Section 6.5 that spontaneously emitted radiation propagates in all directions (in the form of a dipole wave). In real experiments we usually work with light beams with a more or less well defined propagation direction, which can be prepared (unless a laser is used) with the help of diaphragms, i.e. non-transparent screens with a small hole. This direction selection means classically that a piece of the dipole wave is cut out.

At very small intensities the particle nature of the photon again comes into play: an incident photon is either completely absorbed by the screen or passes completely through the hole. The outgoing photon is then characterized, not only by the frequency spectrum (determined mainly by the lifetime of the atomic levels), but also by the propagation direction. A momentum is also related to these two quantities. Formally, the value of the momentum is found by using Einstein's famous equivalence relation between energy E and mass m: $E = mc^2$ (where c is the velocity of light). Thus, we calculate the photon mass to be $h\nu c^{-2}$ and multiply it by the propagation velocity c. The absolute value of the photon momentum is then $h\nu/c$, and, because it points in the propagation direction, we have finally

$$\mathbf{P} = \hbar\mathbf{k}, \qquad (6.11)$$

where $\mathbf{k}$ is the wave vector.

In fact, it is known from classical electrodynamics that the electromagnetic field is characterized not only by an energy but also by a momentum density. Equation (6.11) can be derived from the momentum density by considering, according to the photon concept, a plane wave as being formed from energy packets of magnitude $h\nu$ all propagating in one direction.

The momentum of the photon is experienced by the medium when it is absorbed or reflected by the medium (in the reflection case, twice the momentum is transferred), and this gives rise to the light pressure. The existence of light pressure had previously been anticipated by Kepler in 1617 based on an ingenious interpretation of the observation that comet tails are always directed away from the sun and so it looks as if the particles forming the tail were repelled by the radiation emitted from the sun. Measurements of light pressure on objects on Earth were successfully performed at the beginning of the twentieth century (see, for example, Lebedew (1910)). More recently, light pressure was used to considerably slow down atomic beams (see, for example, Prodan, Phillips and Metcalf (1982)).

The photon momentum is noticeable not only in absorption but also in the process of spontaneous emission. Owing to the validity of the law of momentum conservation, the atom suffers a measurable recoil when it emits a photon, as was predicted by Einstein (1917). Here, it is assumed that the photon is emitted in a well defined, though random, direction (Einstein speaks of "needle radiation;") however, how can this result be brought into accord with the results of Section 6.5, namely that the atom radiates like a Hertzian dipole in all directions, so making possible the interference between partial waves emitted (by the same atom) along different directions? As will be explained in detail in Section 7.5, quantum theory provides the following answer: it is impossible to observe in one experiment both an atomic recoil (indicating the particle nature of light) and interference (which is understood to be a consequence of the wave nature of light).

Besides momentum, the photon also possesses angular momentum – known as spin. This is closely related to the polarization properties of light. It is well known that a light beam can be linearly or circularly (or, more generally, elliptically) polarized. Basically there are two independent polarization states which are conveniently chosen to be either linearly polarized in two mutually orthogonal directions or left and right handed circularly polarized, respectively. These states define an (orthogonal) basis, and any polarization state can be expanded with respect to it. In particular, a circularly polarized wave can be expressed as a superposition of two linearly polarized waves and, vice versa, a linearly polarized wave can be expressed as a superposition of a left handed and a right handed circularly polarized wave.

It is well known that, for a circularly polarized wave, the oscillation direction of the electric field strength, observed at a fixed point, rotates around the propagation direction. Hence we expect intuitively that it has a non-zero angular momentum. A more accurate analysis leads to the statement that the spin (in the propagation direction) of the photons, corresponding to these (classical) waves, takes the possible values

$$s = \pm \hbar. \qquad (6.12)$$

A quantum theorist calls the positive case right handed, the negative case left handed circular polarization. (We emphasize that the classical convention is opposite to this because the sense of rotation is judged by an observer who sees the wave propagating towards himself.)

Similar to the transfer of momentum, the spin of a photon is also transferred to the absorbing medium. The effect is made stronger when circularly polarized light is sent through a $\lambda/2$ plate (a disk consisting of anisotropic, transparent material whose length is chosen in such a way that the path difference between the ordinary and the extraordinary beam equals half a wavelength). It turns left handed circularly polarized light into right handed circularly polarized light, and vice versa, and so each photon transfers an angular momentum of $2\hbar$. The total angular momentum transferred to the plate by many photons could, in fact, be successfully measured using a torsion meter (Beth, 1936).

Let us point out that, within the quantum mechanical description, the circularly polarized states are eigenstates of the photon spin operator (more precisely, of the spin component in the propagation direction). The spin values $\pm\hbar$ are the corresponding eigenvalues (and hence are sharp). This does not, however, apply to linear polarization. In this case the spin (in the propagation direction) is zero *on average*. Because – in analogy to the decomposition of a linearly polarized wave into a right handed and a left handed circularly polarized component – the quantum mechanical state vector of a linearly polarized photon can be written as the sum of two state vectors representing a left handed and a right handed circularly polarized photon, the measurement of the spin component yields the result $+\hbar$ or $-\hbar$, but never zero. Finally, let us mention that the angular momentum conservation law plays an important role in spontaneous emission. Because the atomic levels have defined angular momenta, the conservation law implies certain selection rules for the total angular momentum of the emitted photon (this applies to the absolute value of the angular momentum as well as to the component in the "quantization direction.") However, we have to keep in mind that the total angular momentum is composed of the orbital angular momentum and the spin. The total angular momentum can take values larger than unity (in units of $\hbar$); in this case, we deal with electric or magnetic multipole fields which are observable in the emission of γ quanta by atomic nuclei, for example. However, in optics we deal only with (electric) dipole radiation. In this case the total angular momentum equals unity.

7

Interference

7.1 Beamsplitting

Interference phenomena are certainly among the most exciting phenomena in the whole of physics. In the following we will concentrate mainly on interference of weak fields; i.e. the beams contain, on average, only a few photons.

The principle of classical interference is as follows: a light beam is split by an optical element, for example by a semitransparent mirror or a screen with several very small apertures, into two or more partial beams. These beams will take different paths and are then reunited and form interference patterns. The first step, the splitting of the beam into partial beams, plays a decisive role; light beams coming from different sources (or from different spatial areas of the same source) do not interfere with each other!

We start our discussion of interference with an analysis of the action of a beamsplitter. To form a realistic idea of this device, let us imagine a semitransparent mirror. (Our considerations apply equally well to a screen with two apertures. We could also generalize to cases of unbalanced mirrors, with reflectivity different from 1/2, or screens with apertures of different size.)

The classical wave picture can describe interference phenomena without any great effort: the incoming beam is split into the reflected and the transmitted partial wave, and each of these waves contains half of the energy. The process of splitting becomes conceptually difficult only when we think of the beam as consisting of spatially localized energy packets, or photons. Then, fundamental questions arise. What happens to the individual photon when it hits the mirror? Does it split or does it remain as a "whole?"

There is almost no doubt that photons are – in the sense of energy quanta – indivisible. To understand this we actually need not perform any experiments, it suffices to review all the consequences a divisibility of the photon would imply. Assuming that there is no light frequency change in reflection or refraction, "half photons" (energy packets with an energy content of $\frac{1}{2}h\nu$) would exist in nature. Such objects – when we do not challenge our insight into microscopic processes summarized in Bohr's second postulate, Equation (5.1) – cannot be in any way absorbed again because the energy of a "half photon" is not enough for an atom to make a transition; they would – taking their energy along with them – disappear from the observable world. With the world facing an energy crisis, this is not a very pleasant idea! Even worse, one of the supporting columns of physics, thermodynamics, would be shattered. Let us look at cavity radiation. The insertion of the smallest mirror into the cavity would prevent the formation of thermodynamic equilibrium because the half photons would penetrate the cavity walls and would remain unaffected. The system would therefore lose energy continuously and irreversibly. We should point out at this point that this thermodynamic "catastrophe" would occur already when a fraction (in principle arbitrarily small but finite) of the photons incident on the mirror were split.

The thermodynamic argument can be further refined. Another principle valid in thermodynamics is the principle of detailed balance, which, in the case of cavity radiation, means the following. In thermodynamic equilibrium, the atoms of the walls at each frequency absorb exactly the same amount of energy as they emit (spontaneously and by stimulation). This principle would be violated, however, when the photons split because the half photons no longer contribute to the absorption rate. Even the daring hypothesis that the atom, when not able to absorb half a photon, will have to "swallow" two of them at once, does not improve the situation. Such an absorption process could only take place when at least two half quanta arrive more or less simultaneously at the same position, which is extremely improbable at low intensities. The corresponding transition probability would be proportional to the light intensity squared. The acts of absorption would be too rare to be able to compensate for the decrease (due to the splitting of photons and the associated loss of photons) of the normal (one-photon) absorption rate, with its probability proportional to the intensity.

Finally, let us mention that the splitting, were it real, would not be confined to cutting a photon "in half." First, there exist reflecting surfaces with quite different reflection properties which would cause splits into all possible ratios, and second, a reflection (or transmission) is often followed by a second, third, etc., so that the electromagnetic energy would become increasingly "fuzzy". Photon splitting seems to us to be sufficiently disproved by consideration of these arguments.

Ideal experimental conditions for a direct check of the indivisibility of single photons would provide single photons (sent one after the other) incident on a semi-transparent mirror with the reflected partial beam and the transmitted partial beam monitored by detectors. The response of only one of the detectors would, without any doubt, indicate that the registered photon took one route as an energetic whole. Such an experiment seems to require quite an amount of experimental effort. In fact, measurements with feeble light from conventional sources allow us to make a decision in favor of or against the classical concept of photon splitting via reflection on a semitransparent mirror. In such a situation we have also to expect rare cases when two (or in even rarer cases three or more, etc.) photons arrive more or less simultaneously at the mirror. This corresponds classically to a wave packet with energy $2h\nu$, which is split into equal parts by the mirror. Each part can (with a probability determined by the detection sensitivity) cause a photodetector to respond. However, in most cases, the mirror is hit by at most one photon within a time interval of the order of the response time of the detector. Such single photons would be – if split by the mirror – missed by the detectors, which are able to handle only whole quanta. On the other hand, a measurement on the incident beam would detect them (according to the sensitivity detection). This would imply a drastic violation of the energy balance

$$\text{incident intensity} = \text{reflected intensity} + \text{transmitted intensity} \qquad (7.1)$$

as long as photodetectors are used for the measurement. Equation (7.1), which can be understood only in such a way that the photons as a whole are either reflected or transmitted, is fully verified by experiment (Jánossy, 1973).

On analyzing all the experiments performed to check the indivisibility of the photon, we should keep in mind that, due to the finite measurement precision, we cannot completely exclude a possible splitting of the photon, though this would have a very small probability. Indeed, an experiment can always reveal only a finite upper bound but never an exact zero value. Hence, we prefer the thermodynamic argument, which is free of the experimental insufficiencies; we see it at least as a not unimportant supplement to the experimental facts.

We now take the problem of beamsplitting a few steps further. For large intensities of the incident beam, the following question arises. How are n (> 1) incident photons packed in a short pulse[1] divided between the two partial beams?

A naive way of treating the problem of the interaction between light and a beamsplitter would be to resolve (at least theoretically) the process into a sequence of independent individual processes which always involve just a single photon. Assuming, in addition, the photons to be classically distinguishable particles, we can apply classical probability theory and present the following argument. A mirror with reflectivity r (the mirror is ideal – no losses whatsoever) and transmittivity

---

[1] In principle, it is possible to prepare such a light "flash" of a sharp photon number by bringing a certain number of excited atoms into a small volume and taking care, with the aid of concave mirrors, that the outgoing radiation is directional (Mandel, 1976).

$t = 1 - r$, reflects a photon with probability r and transmits a photon with probability t. The probabilities of $k(\leq n)$ photons being transmitted and $n - k$ photons being reflected, out of a total of n incident photons, is given by

$$w_k^{(n)} = \binom{n}{k} t^k r^{n-k} \quad (k = 0, 1, 2, 3, \ldots, n). \tag{7.2}$$

The first factor accounts for the fact that the event – due to the possibility of distinguishing between the photons – can be realized in different ways. (While an interchange among reflected or transmitted photons does not lead to a new case, the interchange between a transmitted and a reflected photon does.)

Equation (7.2) is a binomial distribution for the photons. Since it is very difficult to prepare n-photon states with $n > 1$, an experimental verification is presently limited to the case of photon pairs (Brendel *et al.*, 1988), which can be quite easily produced using parametric fluorescence. However, the assumptions used for deriving Equation (7.2) are wrong: since the photons arrive at the mirror more or less simultaneously, the mirror "feels" (due to the superposition principle of classical electrodynamics) their resulting electric field, and so they act together. In addition, because they can be viewed as particles of integer spin, they follow Bose statistics, and are therefore indistinguishable. Surprisingly, the quantum mechanical calculation also leads to Equation (7.2) (see Section 15.4), which therefore can be assumed to be correct.

The above derivation allows us to say that the photons, in the process of beam-splitting, behave *as if* they were distinguishable and *as if* each of them interacts independently with the mirror. However, it would be erroneous to try to objectify these properties. When the number of photons is not sharp but follows a probability distribution p_n, the number of transmitted photons will be given (in accordance with Equation (7.2)) by the distribution

$$p_k' = \sum_n \binom{n}{k} t^k (1 - t)^{n-k} p_n. \tag{7.3}$$

(An analogous relation holds for the reflected photons.) Equation (7.3) is known as the Bernoulli transformation. It has the important property that the factorial moments are simply multiplied by the respective power of t; i.e. the following relation holds:

$$\overline{k(k - 1) \cdots (k - l)} = t^{l+1} \langle n(n - 1) \cdots (n - l) \rangle \quad (l = 0, 1, 2, \ldots). \tag{7.4}$$

Applying Equation (7.3) to a Poisson distribution of the incident photons over the mode volume (as explained in Section 4.2, we identify it with the pulse or coherence volume, respectively), we find that the reflected and the transmitted photons follow a Poisson distribution. The result is of particular physical interest because the Poisson distribution, as described in Section 4.4, is characteristic of (quantum mechanical) coherent states of the electromagnetic field. The result indicates that beamsplitting transforms a coherent state back into a coherent state,[2] and this is what we expect from the correspondence between the classical and the quantum mechanical description. Coherent states correspond to classical waves of sharp amplitude and phase, and, according to classical optics, the transmitted as well as the reflected wave have sharp values of amplitude and phase when the same applies to the incident wave.

Let us mention that the splitting of a polarized light beam into two beams with different polarizations (for example a linearly polarized light wave, with the help of a birefringent crystal, is split into two waves, mutually orthogonally polarized in different directions) is of the same kind as beamsplitting with a partially transmitting mirror. Equations (7.2) and (7.3) apply also in this case. Another consequence of Equation (7.3) is that thermal light remains thermal after splitting. This corresponds to our expectations. Finally, let us point out that the partially transmitting mirror can serve as a model of a (one-photon) absorber; in fact, the quantum mechanical description is, in both cases, formally identical. We will use this fact in Chapters 8 and 10.

7.2 Self-interference of photons

All known interference phenomena can be naturally explained by representing light as waves (see Section 3.2). Let us demonstrate the interference effect using the example of a Michelson interferometer (see Fig. 7.1). An incident light beam is split by a semitransparent mirror into two partial beams. The beams propagate along the interferometer arms, are reversed by mirrors and are finally reunited by the semitransparent mirror, whereupon they enter the observation telescope. Because the mirrors at the ends of the arms are usually not exactly orthogonal to the light beams, we observe – similar to the case of a wedge – interference fringes of equal thickness.

How do we understand the formation of an interference pattern in the photon picture? As discussed in detail in Section 7.1, we have to accept that the photons (in the sense of energy packets) are not split by the semitransparent mirror – in contrast to wave packets. It seems that the photons can take only one of the two paths and hence can "know" the length of only one of the interferometer arms. On the other hand, the position of the interference fringes is determined by the

---

[2] Coherent states are represented by a wave function (see Equation (4.5)), and it is not sufficient, as has been done up to now, to characterize them only by the squared absolute values of the expansion coefficients (in the photon number basis) determining the photon distribution. Only with the help of the transformation, describing beamsplitting, of the expansion coefficients can the above conclusion be rigorously drawn (see Section 15.4).

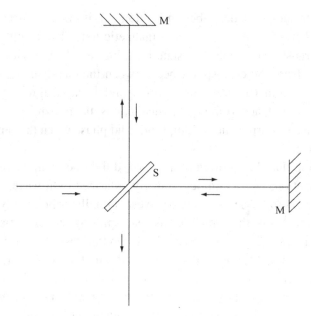

Fig. 7.1. Beam paths in a Michelson interferometer. S = semitransparent mirror;
M = totally reflecting mirror.

length *difference* between the two arms. We might conjecture that several photons
must necessarily cooperate to produce interference. This would however imply a
drastic intensity dependence of the interference effect: with decreasing intensity,
the interference pattern should become more and more blurred because a photon
taking one of the interferometer arms would rarely find a second photon taking
the other path. Finally, in the ideal case of single photons incident one after the
other, there should be no interference, which means that the black spots on the
photographic plate should be distributed completely at random.

In contrast, in the classical picture, wave splitting into partial waves is completely independent of the intensity and always takes place in the same way; i.e.
according to classical optics, the visibility of the interference pattern

$$s = \frac{I_{max} - I_{min}}{I_{max} + I_{min}}, \tag{7.5}$$

where I_{max} is the maximum value and I_{min} is the minimum value of the intensity
distribution in the interference pattern, is the same for all intensities.

What do we learn from the experiment? Interference experiments with extremely weak intensities had already been performed early in the twentieth century (Taylor, 1909; Dempster and Batho, 1927). To see anything at all on the photographic plate under such conditions, extremely long exposure times had to be
chosen. The longest measurement in the classic experiment by Taylor (1909)

took three months! There was no appreciable deterioration in the interference pattern visibility, and hence the wave theory was proved to be correct. More recent experiments using photoelectric detection methods confirmed this result (Jánossy and Náray, 1957, 1958; Reynolds, Spartalian and Scarl, 1969; Grishaev *et al.*, 1972; Sillitoe, 1972).

Jánossy and Náray (1958) (compare also Jánossy (1973)) also analyzed in their experiments the question of whether the length of the (optical) paths that the partial beams have traversed before they become reunited has any influence on the fringe visibility. The authors used a Michelson interferometer with an arm length of 14.5 m, which was significantly larger than the coherence length of the analyzed light. The apparatus was placed in tunneled-out rock, 30 m below the surface, to guarantee the required stability of the experimental setup. The measurements were fully automated to avoid disturbing the thermal equilibrium by human observers, since temperature fluctuations of just 0.001 K caused a shift in the interference pattern. The authors were disappointed because, even for strongly attenuated incident light intensities, no fringe visibility deterioration could be observed. (To appreciate fully the experimental accomplishment, we have to understand that it is a simple matter to cause the disappearance of the interference pattern – all that is required is to take no care when performing the experiment!)

Another original experiment was carried out by Grishaev *et al.* (1972), who analyzed the ability of synchrotron radiation, generated by electrons orbiting in a storage ring, to interfere. The measurement of the intensity of the emitted radiation was used to determine the number of electrons stored in the ring, and this allowed the registration of an interference pattern for a fixed number of electrons N. It turned out that the visibility of the interference pattern remains the same for N varying between 16 and 1. In the case of a single electron, about 1.3×10^5 photons entered the interferometer per second, which implies a mean temporal distance of 0.8×10^{-5} s, orders of magnitude larger than the interferometer passage time of 10^{-9} s.

Finally, let us mention that time domain interference (the beats mentioned in Section 3.2) for low intensities, as in the aforementioned experiment, has been observed (Davis, 1979).

Dirac's statement that "a photon interferes with itself" (Dirac, 1958) can therefore be considered to be experimentally proven. This fact is completely incomprehensible when the photon picture – in the form of localized energy packets – is made objective in accordance with the classical concepts of reality. This is, however, forbidden by quantum mechanics! It is forbidden (by quantum mechanics) to draw conclusions about certain physical properties *before* the measurement of those properties is complete. In the case of beamsplitting, this means that it is possible to conclude from the click of a counter detecting a photon as a "particle" in one of the partial beams that the corresponding path was taken, but not that this would have been so in the absence of a counter. Indeed, the behavior of the photon must change considerably, as the observable "interference with itself" proves, when the beamsplitter is put together with other devices to form a Michelson interferometer. The experimentally "posed question" is then completely different: we are not at all interested in the path the photon has "really" taken, but in the precise point of its arrival at the focal plane of the telescope. Under such circumstances the wave picture is the proper one (until the act of photon detection).

We also have to accept in this case the wave–particle duality of light as a matter of fact: depending on the experimental conditions, the particle or the wave nature of light manifests itself. Light is neither a particle nor a wave; it is something more complicated, which sometimes shows its particle side and sometimes shows its wave side.

Quantum mechanics succeeded in synthesizing these two contradicting aspects, but the price to pay was the classical reality concept. Quantum mechanics describes the state of the total field consisting of a transmitted wave and a reflected wave (in the following we label the two waves 1 and 2) by a wave function, which in the case of a single incident photon is a superposition,

$$|\psi\rangle = \frac{1}{\sqrt{2}}(|1\rangle_1|0\rangle_2 - |0\rangle_1|1\rangle_2), \tag{7.6}$$

of the states "the photon is in beam 1 and not in beam 2" and "the photon is in beam 2 and not in beam 1" (see Section 15.4). This means in particular that the assumption that the photon is *either* in the one *or* in the other beam is wrong (in such a case there would be no interference), and so we face a fundamental uncertainty (going far beyond simple ignorance and hence classically not interpretable) about the path of the photon. The minus sign in Equation (7.6) takes into account the reflection induced phase change of π. The casual saying, the photon is both in one partial beam as well as in the other partial beam, comes closest to reality: when the photon wants to interfere with itself, it must somehow "find out" the distances between *both* mirrors and the beamsplitter (see the Michelson interferometer in Fig. 7.1).

We cannot envisage, however, the simultaneous "presence" of the photon in both beams as a split of the photon energy between the two beams where it would be objectively (in the sense of a classical description of nature) localized. In such a case we run into a dilemma when trying to describe the experiment complementary to interference, the (photoelectric) detection of the photon in one of the two beams. The photodetector responds within a very short time (see Section 5.2), and, due to the finite propagation velocity of the electromagnetic energy, there would be not enough time left to "get back" the remaining required energy from the other beam, at least not when the corresponding path is too long. The path is not subjected to any fundamental limitation; in practice, extremely long paths can be achieved using optical fibers, avoiding divergence of the light bundle (as it takes place in the vacuum).

We come to the same conclusion as we did for the case of spontaneous emission (Section 6.3): the classical concept of continuously distributed electromagnetic energy (according to the values of the electric and the magnetic field strengths) in space must be abandoned in the case of single photons.

It is important to note that the statement regarding a self-interfering photon must not be taken literally in the sense of being verifiable with a single measurement. If we consider the detection of a possible interference pattern using a photographic plate, we see that a single photon delivers a single black spot on the plate which can equally well be an "element" of an interference pattern or a member of a completely random distribution of dark spots.

Only in the case where the photon produces a black spot at a position for which an exact zero of the intensity is predicted can a single measurement provide definitive conclusions regarding the interference properties of the photon, namely that the predicted interference pattern does not agree with the facts. (To falsify a statement, we only need a single contradicting experiment.) In reality, such cases are of academic importance only because the ever present imperfections (such as the finite bandwidth of radiation, deviations in the reflectivity of the semitransparent mirror from $1/2$, etc.) hinder the minima of the intensity in reaching zero.

As always, to verify quantum mechanical statements we have to perform many individual experiments (under identical physical conditions). In the case of interference, the black spots obtained in this way all together form the interference pattern (photographed one on top of the other). It is important that the time difference between two individual experiments can be chosen to be arbitrarily large so that each of the experiments involves only a single photon. Hence, the appearance of an interference pattern is always observable only on an ensemble of (independent) individual systems.

On the other hand, an interference pattern can obviously be formed only when the photons follow certain "rules." These rules state that certain positions (the areas where interference maxima are formed) are preferable "addresses" for the photons to reside at, while other positions (corresponding to minima) are avoided. From this perspective, the photon indeed interferes with itself, but we recognize this specific property of a single photon only by the behavior of a group of photons.

Let us note that the quantized theory of the electromagnetic field encompasses the particle equally as well as the wave aspect. In particular, beamsplitting can be described in such a way that (in complete correspondence to classical theory) the electric field strength – now described by an operator – of the incident wave is decomposed into parts corresponding to the reflected wave and the transmitted wave (see Section 15.4). We find then the surprising (at least at first glance) result that the classical interference pattern is quantum mechanically exactly reproducible independent of the (perhaps even non-classical) state of the incident light (see Section 15.5).

We should point out that the validity of the classical description of interference in the domain of microscopically small intensities means that conventional spectrometers, all using the interference principle, are functioning even when a single photon is incident. The photon will be spectrally decomposed; i.e. as a wave it is split into various partial waves corresponding to different frequencies. The measurement can be considered complete only when a result is indicated. For this purpose the outgoing light must be registered. The photon will be found in one of the partial beams, and we measure a *certain* value of the frequency (with a precision defined by the resolution of the spectrometer). Repeating the experiment frequently under identical conditions gives us a frequency spectrum.

Finally, it is appropriate to comment on the concept (mainly attributable to Louis de Broglie and Albert Einstein) of a guide wave for the photon – conceived as a localized energy packet. Some authors hoped to make the interference behavior of the photon comprehensible by invoking this theory. According to the concept, the photon would "in reality" (we might think of a Michelson interferometer, for example) take only one path and just the "guide wave" would be split by the semi-transparent mirror. One part of the "guide wave" would travel along one arm together with the photon, while the other would have to travel alone along the other arm. After reunification of the two parts, the "guide wave" would possess enough information to direct the photon to a position in agreement with the classical description of interference.

In this picture, the "guide wave" behaves exactly as an electromagnetic field in classical theory. The existence of the electromagnetic field is beyond any doubts – at least when many photons are simultaneously present – and hence there is, in our opinion, no reason to introduce a new physical quality in the form of a "guide wave." Although we satisfy our desire to objectify the motion of the photon in space, the price to pay is, as Einstein put it, a world inhabited by "ghost fields." Each beamsplitting without later reunification would lead (in the case of a single incident photon) to a guide wave "being without a job." The wave would split again at a second beamsplitter, etc. and all these "jobless" guide waves would "haunt" us until the news about the "death" of their lost protégé ended this horror!

7.3 Delayed choice experiments

As already mentioned in Section 7.2, we can build an interferometer by supplementing the beamsplitter with additional mirrors. On doing so, the photon behaves as a wave and splits into two equal parts, whereas on direct observation of the process of beamsplitting we see that the particle remains indivisible. We can conclude that the photon behaves as either a particle or a wave depending on the particular experimental setup used.

We can "cheat" the photon by delaying the decision on the choice of experimental setup until after it has passed through the beamsplitter. If the photon found at its arrival at the beamsplitter a situation requiring it to "reveal" its particle properties, it could confidently "choose" to take one of the paths. But what would happen if, after a time, it were asked to participate in the formation of an interference pattern, and hence to take both paths?

Admittedly, such a point of view is rather naive – the photon would have to be equipped with clairvoyant abilities to be able to obtain the necessary information about the *complete* experimental setup at the (first) beamsplitter because its field distribution at this time would be unable to "feel" the additional mirrors, which, in principle, could have been placed arbitrarily far away. Nevertheless, the problem stimulated several experimentalists to construct so-called "delayed choice experiments." We will briefly discuss one of them below (Hellmuth *et al.*, 1987).

The experiment was based on a Mach–Zehnder interferometer setup (see Fig. 7.2). The appearance of interference is detected by the different light intensities at the outputs. The actual intensity ratio between the outputs depends on the setting of the interferometer (the difference of the interferometer arm lengths). In particular, the interferometer can be tuned so that all the light leaves through the same output. The subsequent single incident photons are detected in one or other of the outputs; the different frequencies of these two kinds of events indicate the interference of the photon with itself.

The details of the particular experiment were as follows (Fig. 7.2). A good approximation of a single photon source was achieved by strongly attenuating pulses of 150 ps duration, so that the mean photon number per pulse was only 0.2. A Pockels cell was placed in the upper arm of the interferometer, forming, together with a polarizing prism, an electro-optical switch. On applying a voltage to the Pockels cell, birefringence was induced, and the polarization direction of the incident (linearly polarized) light was rotated by 90° and was thus deflected by the prism from the interferometer; the interferometer arm was blocked. When the voltage was removed the polarization rotation did not take place and the light passed through the polarizer without changing its direction. The Pockels cell was operated in such a way that the upper interferometer arm was blocked and was opened only after

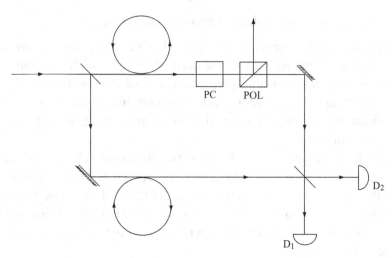

Fig. 7.2. Delayed choice experiment. PC = Pockels cell; POL = polarizer; D1, D2 = detectors.

the photon had passed through the beamsplitter at the entrance. More precisely, the photon was in the glass fiber between the beamsplitter at the entrance and the Pockels cell. Delay lines in the form of a 5 m long single-mode optical fibre had been inserted into the optical paths of both interferometer arms to allow enough time for the Pockels cell to be switched off. This device presented the photon with the dilemma discussed at the beginning of this section, but the experiment showed that the photon was not impressed by it in any way: it chose its behavior only in accordance with the experimental conditions it found at its moment of arrival at the given place, and identical interference patterns were observed independent of whether the upper interferometer arm was always open or only opened when the photon had just passed through the beamsplitter.

7.4 Interference of independent photons

It is a well known fact in optics that light waves emitted from different light sources (or from different points of the same source) cannot be made to interfere. The reason can be traced back to fluctuations of the phase of the electric field strength. (The emitted light is formed by contributions from individual independently emitting atoms – which therefore have statistically random phases – and, because the elementary emission process, as described in Section 6.2, has a short duration, both the phase and the amplitude of the total field changes in an uncontrollable manner.)

The phase fluctuations do not matter in the standard interference experiment (in which partial waves originating from the same primary beam interfere) because the phase changes in the partial beams are exact copies of the phase fluctuations of the master beam. Consequently, the *relative phase* between two different beams determining the position of the interference pattern according to Equation (3.14) remains unaffected by the phase fluctuations and is determined only by the geometry of the setup. Only when the path difference between the partial waves exceeds the coherence length of the used light, the phases do not generally "match" each other, because at a given time the (individual) phases have on average an approximately constant value only over distances of the order of the coherence length; over longer distances the phase changes randomly. Thus, in the case of a Michelson interferometer, the interference pattern vanishes when the arm difference exceeds the coherence length.

For independently generated optical beams the phases fluctuate in a completely uncorrelated way and the interference pattern is shifted perpetually by random amounts so that it is completely "washed out."

So, can we say that the reason we do not observe an interference pattern in the experiment is the large observation time? However, because the individual phases observed at a fixed point do not change significantly within times of the order of the coherence time (by this we mean the coherence length divided by the velocity of light), an interference pattern should be detectable at a time scale of the order of the coherence time.

Indeed, it is not too difficult to fulfil the above requirement in the present experimental state of the art. For this purpose an electro-optical shutter or an image converter which is gated on for a short time by an electric control pulse could be used.

This is not the end of the story, however. The number of photons must be great enough to be able to draw definitive conclusions about an interference pattern from the blackened spots on a photographic plate (or from photoelectrically obtained measurement data). It turns out that the requirement cannot be met in practice with thermal light sources (gas discharge lamps, etc.). Let us emphasize, however, that there are no obstacles in principle. According to Planck's radiation law, the spectral energy density, and with it also the number of photons incident on a detector within the coherence time, increases with increasing temperature. However, for the discussed interference experiment, the required temperature would be unrealistically high!

There is still hope, however, as we can turn to novel light sources – lasers – which can deliver light with fantastically high spectral densities. Spatial interference between two light pulses – with slightly different propagation directions – was observed in the early 1960s (Magyar and Mandel, 1963). The pulses were emitted

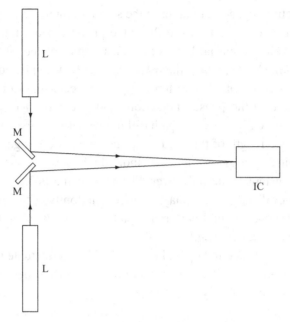

Fig. 7.3. Interference of two laser beams. IC = image converter; L = laser; M = mirror. From Magyar and Mandel (1963).

in an irregular sequence (in the form of so-called "spikes") by two independent ruby lasers (Fig. 7.3). Because the phases of the laser pulses change randomly, the interference pattern changes its position from shot to shot. The high intensity of the laser light created sufficiently many photoelectrons in each run – Magyar and Mandel used an image converter, which was electronically gated on for a short time only when both lasers simultaneously emitted a pulse – such that an interference pattern was formed.

A year before this, quantum beats between two laser waves with slightly differ-ent frequencies had been demonstrated (Javan, Ballik and Bond, 1962). This ex-periment is easier to achieve because one can work in the continuous regime. The photocurrent of a photomultiplier follows the time evolution of the total intensity of light and hence contains, according to Equation (3.14), a contribution oscillating at the difference frequency of the lasers, and this represents the beat signal.

Beats are very easy to observe using gas lasers because their frequencies – in each of the excited eigenoscillations or "modes" – are very sharp (the linewidth is of the order of 10^{-3} Hz). Under normal working conditions, and over rather long time scales, we can observe frequency shifts caused primarily by mechanical instabilities of the resonator setup (they are typically of the order of 100 kHz/s). Precisely this frequency "drift" can be monitored in beat measurements with high accuracy, which was in fact the primary aim of the experiment.

Interestingly enough, such a beat experiment was performed before the laser era (Forrester *et al.*, 1955), and we do not want to miss the opportunity to comment in some detail on this pioneering work. Not only did the authors have to master enormous experimental problems, but they also had to counter the doubts presented by theorists that the desired interference effect was even possible.

A microwave excited electrodeless gas discharge in a tube filled with the mercury isotope Hg^{202} was used as a light source. The green mercury line at a wavelength of 546.1 nm was chosen for the measurement, and it was split into different components by an external magnetic field, so exploiting the Zeeman effect. The components are, when observed in the direction orthogonal to the magnetic field, linearly polarized either parallel to the magnetic field (π component) or orthogonal to it (σ component). The authors intended to detect beats between two components (polarized, of course, in the same way).

At the rather modest intensities Forrester *et al.* had to work with, shot noise proved to be the biggest obstacle to observation. (Because white shot noise was present, there was also a noise component oscillating at the beat frequency.) During a beat period, however, enough photons impinged onto the cathode (about 2×10^4) to allow the formation of the beat signal. However, the experimental conditions were such that the phases of the light waves (at each moment) were constant only over very small areas of the illuminated cathode surface, the so-called coherence areas, and changed from one area to another in a completely random way. Hence, the contributions from the individual coherence areas to the total alternating current of the photocathode, i.e. to the beat signal, join together with random phases, making the signal almost vanish in comparison to the shot noise (originating from the direct current part of the photocurrent).

The estimation of the signal to noise ratio produced a value of about 10^{-4}. In such a situation, only "marking" the signal could help. The authors used a well known technique: they modulated the signal in its intensity (at an unchanged noise level) and then amplified it with a phase sensitive narrowband amplifier to distinguish it from the noise. The required modulation was accomplished with the help of a rotating $\lambda/2$ plate with a polarization foil positioned behind it. The rotating plate caused a rotation of the polarization direction of the two σ components whose interference was to be measured, the speed of rotation being twice the speed of the plate. The polarization foil, which transmits only light of a certain polarization, converted the rotation of the polarization into a periodic intensity modulation. The other σ components and the π components were also influenced; however, the intensities of the π components are minimal when the σ component intensities become maximal. When the light of the analyzed spectral line is *in total* unpolarized (to achieve this polarization already present had to be compensated for) the total intensity of the spectral line impinging on the photocathode, and hence also the receiver noise, remains constant in time, as required.

Using the described modulation technique, Forrester *et al.* managed to enhance the signal to noise ratio from 10^{-4} to a value of 2. However, even under these circumstances, in the authors' words, "a great amount of patience was needed to obtain data." The observed beat frequency had to be chosen to be relatively high to be able to attain the required separation of the Doppler broadened Zeeman components. In the experiment it was 10^{10} Hz, i.e. in the microwave region. Therefore, a microwave resonator was used to measure the beat signal excited by the electrons coming from the photocathode. From the agreement between the measurement data and the result of calculations performed by them, Forrester *et al.* concluded that they had succeeded in observing the interference.

Let us now return to spatial interference which demonstrates so clearly the interference phenomenon. As previously explained, the interference between two intense independent light beams is an experimental fact. For radio engineers this comes as no surprise because in the early days of radio communication they quickly learnt that radio waves coming from different sources have the unfortunate property that they interfere. Researchers in optics have had to wait to find an analog to the radio transmitter, and this wait ended with the invention of the laser.

What happens to the ability of the waves to interfere at very low intensities? Do laser beams interfere after they are strongly attenuated? At first glance it seems that the answers are no, for fundamental reasons. Laser beams (we mean continuous laser operation) have a finite linewidth and hence a defined coherence length. When the intensity is decreased to such an extent that the coherence volume (i.e. a cylinder with its base equal to the beam cross section and its height equal to the coherence length) contains only a few photons, we run into the same difficulties as in the case of thermal radiation. In contrast to this, however, there is now a way out of the dilemma. The intense light beams, from which we "split off" a tiny part for the interference experiment, can be used to obtain information about the phase of our *attenuated* beam. Having this information at hand, we can control a shutter so that the photoelectrons are detected only when the phase between the two interfering beams has a prescribed value. The experimental setup would basically have the following form (Paul, Brunner and Richter, 1965; Paul, 1966). Light waves emitted from two identical lasers each impinge on a beamsplitter which reflects only a small fraction of the incident radiation (Fig. 7.4). The reflected beams (being only slightly different in their directions of propagation) – after being further attenuated by an absorber when necessary – are made to interfere on an observation screen S_1. Before this, however, they pass an electronically controlled shutter which is opened only when the momentary interference pattern on screen S_2 generated by

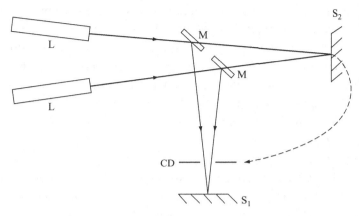

Fig. 7.4. Experimental setup for the observation of interference between small numbers of independent photons. CD = controllable diaphragm; L = laser; S_1, S_2 = screens; M = weakly reflecting mirror.

the transmitted (intense) laser beams has the prescribed position. In this way it is also possible to detect interference patterns for very low intensities – the exposure time must be just long enough.

The described experiment was successfully achieved in the 1970s by Radloff (1971). (Radloff had earlier observed beats between two independent strongly attenuated laser beams (Radloff, 1968).) The interference was observed for beams generated by two He–Ne lasers operated in the single-mode regime. Good mechanical stability was achieved by using a cylindrical quartz block with two longitudinal drills. To one face of the quartz block were fastened two ceramic hollow cylinders (for the purpose of piezoelectric length change). On the ends and the other face of the quartz block were the laser mirrors. Both laser tubes were inserted into lateral slots of the quartz block. To obtain the control signal for the diaphragm (formed by an electro-optical shutter) the beat signal between the two (intense) laser beams was detected.[3] The procedure was such that when the beat frequency was within the range 3 to 70 kHz, the beat maxima were amplified and converted into rectangular pulses, which opened the electro-optical shutter. The mean photon current in the attenuated laser beams was 10^5 photons/s. The opening time of the shutter varied between 10^{-4} and 10^{-5} s so that only a very few photons arrived at the photographic plate. Nevertheless, in this case – for a measurement time of 30 min – a well defined interference pattern could be observed.

---

[3] As already noted, the finite linewidth of the laser radiation (for the gas laser) is due to the long-term frequency drift originating from the mechanical instability of the resonator. The drift leads to a frequency difference between the two lasers, and this is the main reason for the displacement of the interference pattern. Observing the time evolution of the interference at a fixed position it appears as a beat, and obviously the interference pattern always has the same position when the beat amplitude takes the same value (for example its maximum).

In fact, a quantum mechanical description of this interference phenomenon (see Section 15.5) leaves no doubts that the visibility of the interference pattern is independent of the intensity. Interference appears for (in principle) *arbitrarily small* intensities. The *mean* number of photons reaching screen S_1 within the coherence time can, in fact, be much smaller than one! Then, during most of the time intervals when the diaphragm is opened, nothing happens, but, from time to time, a photon is registered that – like an individual piece of a mosaic – contributes to the interference pattern.

Instead of making the interference pattern directly visible in this way, we can "reconstruct" it afterwards. To do this we dispense with the diaphragm, which naturally causes the blackened spots on screen S_1 to be distributed completely at random, i.e. no interference can be recognized. For each blackened spot an observer[4] registers the time of registration, while a second observer, using the interference pattern formed by the intense laser beams, registers the relative phase between the two waves as a function of time. After finishing the experiment the first observer can still "recover" the interference pattern when the measurement data of the second observer are available. It just remains to select those blackened spots which correspond to moments when the relative phase had the same (prescribed) value.

Let us note, by the way, that it is not necessary for the observers to make simultaneous measurements. Using a sufficiently long delay line (a glass fiber for example) the second observer can, at least in principle, perform observations after the first observer has finished taking measurements. In this case, the reconstruction of the interference pattern after both measurements are completed is possible; however, we must know the exact time spent by the light in the delay line to be able to count back reliably. All of this is not very surprising because the interference between two intense laser beams represents a macroscopic process and the classical concept of reality applies. This means that we can ascribe a definite position to the interference pattern in the sense of an objective fact, independent of whether (and when) we measure it or not. A possible measurement is then nothing more than taking notice of a previously existing fact. Thus we can also equally well reverse the time order of the measurements performed by the two observers.

How should we imagine the interference between independent photons? Obviously the photon concept completely fails. Because the position of the interference pattern is determined through parameters of *both* beams (propagation direction and phase while assuming coinciding frequencies and polarization directions) we would have to assume that always one photon (at least) from one beam is cooperating some way with (at least) one photon from the other beam. However,

---

[4] The term observer is understood quite generally to include automated measurement devices storing data (such as magnetic tape).

this is extremely difficult to imagine, even for higher intensities, when we are dealing with spatially localized light particles. At very small intensities this is simply
not possible because it is extremely rare that a photon from one beam and, simultaneously, a photon from the other beam pass the shutter during its opening time.
We can illustrate our considerations using a drastic example. Let us assume that
during the opening time of the shutter each laser sends, on average, $1/1000$ of a
photon through the diaphragm, then only in two cases out of 1000 does *one* photon (either from the first or the second beam) hit the photographic plate, while only
once in 10^6 cases does the desired coincidence take place.

We have to resort to the wave picture! The discussed interference is then immediately understandable within the framework of classical electrodynamics, as
explained in Section 3.2, and the intensity does not matter at all. It becomes prob-
lematic again only when we try – for the discussed small intensities – to bring the
wave picture into harmony with the quantization of the radiation energy. Indeed,
all the problems with comprehending this situation arises from the inadmissibility
of making the photon concept "objective."

The concept that each wave pulse has in it *in all cases* a well-defined photon
number (taking, for example, the values 0, 1, 2, etc.) is unjustified according to
the insight provided by quantum mechanics. Let us confine ourselves to the case
when the photon number can take either the value zero or one; quantum mechanics,
apart from the states "there are certainly no photons present" (represented by the
symbol $|0\rangle$) and "there is exactly one photon present" (represented by $|1\rangle$) allows
for a whole range of possibilities, namely the (arbitrary) superpositions $a|0\rangle +
b|1\rangle$ of the two states (a and b are complex numbers satisfying the normalization
condition $|a|^2 + |b|^2 = 1$). In particular, the states can be of such a kind that the
probability $|b|^2$ is arbitrarily small. Nevertheless, such states are fundamentally
different from the corresponding mixture describing an ensemble, the elements of
which are *either* in state $|0\rangle$ *or* – with the correspondingly small probability – in
state $|1\rangle$. For example, the phase of the electric field strength is, in a superposition
state, more or less well defined, whereas for states with a sharp photon number –
and hence also for a mixture of such states – it is completely uncertain (compare
with Section 4.3).

Indeed, the quantum mechanical description of interference between independent photons is based on the assumption that the two light beams are each in a
superposition state of the aforementioned type. In particular, it was proven (Paul,
Brunner and Richter, 1963) that coherent states of the electromagnetic field,[5] as
mentioned in Section 4.4, give rise to an interference pattern which – independent
of the intensity – corresponds *exactly* to the predictions of classical theory using the

---

[5] For very small mean photon numbers ($|\alpha|^2 \ll 1$) the "admixture" of more-photon states $|2\rangle$, $|3\rangle$, etc. does not
practically play any role.

concept of waves having definite phases and amplitudes (see Section 15.5). Actually, laser light can be described, as discussed in Section 4.4, to a very good approximation by coherent states, even after it has been (in principle) arbitrarily strongly attenuated.

Let us look into the special case of an extremely small intensity so that the coherent state in Equation (4.5) can be approximated by $|\alpha\rangle \approx |0\rangle + \alpha|1\rangle$, and the direct product of the two interfering beams can be written in the form

$$|\alpha_1\rangle_1|\alpha_2\rangle_2 \approx |0\rangle_1|0\rangle_2 + \alpha_1|1\rangle_1|0\rangle_2 + \alpha_2|0\rangle_1|1\rangle_2. \tag{7.7}$$

The first term does not play any role in describing the response of the detector – the detector does not react to the vacuum state of the field. The term relevant to the interference has exactly the same structure as that in the case of self-interference of the photon (see Equation (7.6)). The position of the interference pattern is determined by the relative phase of the two complex numbers α_1 and α_2. (Equation (7.7) is more general than Equation (7.6) because the intensities of the two beams can also be different.) As far as the formal description is concerned, there is no difference between the self-interference of a photon and the interference of two independent photons! However, we have to emphasize that the photons "admitted" to interference are not statistically independent. They are, in reality, phase correlated – more precisely, this is valid for the corresponding fields – and because only one photon is ever detected, the physical conditions are indeed quite similar to those present in the case of the interference of a photon with itself. The difference is only evident in the way in which the required phase correlations are generated: conventional experiments use beamsplitting for this purpose, while in the case of independent laser beams the above described "preparation" through the technical arrangement of the measurement selects proper "pieces" from the total radiation field.

The almost ideal correspondence between the quantum mechanical and classical descriptions shows that classical electrodynamics comes much closer to quantum mechanics than classical mechanics, which is able to reproduce quantum mechanical predictions only approximately at best. The reason for this is that electrodynamics is already a *wave theory* and, within its area of applicability – including interference phenomena – is able to perform equally well as wave mechanics or, in other words, quantum theory. This fact may seen surprising especially for a physicist having preconceptions of quantum mechanics. Someone like this can be easily led (and this relates to the author's own experience) into the dark by the quantum mechanical uncertainty relation for the phase φ and the photon number n already derived by Dirac (1927) (compare with Heitler (1954)):

$$\Delta n\,\Delta\varphi \geq \frac{1}{2}. \tag{7.8}$$

Since a prerequisite for the formation of a well visible interference pattern is the presence of sharp phase values of the (independent) partial beams, one is tempted to draw from Equation (7.8) the conclusion that, with decreasing mean photon number and consequently decreasing dispersion Δn, the phases should fluctuate more strongly and thus wash out the interference pattern more and more. This erroneous conclusion is based on the fact that the quantum mechanical phase (defined by the corresponding operator) contains a contribution representing the vacuum fluctuations of the electromagnetic field mentioned in Section 4.3. However, a photodetector does not take any notice of them. The result is that the interference pattern loses none of its visibility in the correct quantum mechanical description, even for smaller and smaller mean photon numbers.

Let us return, after this short detour, to the question of how to imagine the interference between independent photons. From the above discussion it follows that it is uncertain *in principle* how many photons (if any) are in each of the beams. (The physical situation is in this respect basically different from that of a photon interfering with itself. In that case we could safely assume we were dealing with a single photon.) When a photon is detected on the observation screen, it has obviously come from the total field formed by the superposition of the two light waves. The "localization probability" of the photon (on the screen) follows the maxima and minima of the intensity of the superposed field. It is impossible (in principle) to "read out" from which laser the detected photon originated! Photons are not individuals with a "curriculum vitae" that can be traced back! What happens in the experiment is simply nothing more than an act by the detector of taking the energy $h\nu$ from the field, and the question of where it came from is already physically meaningless in the classical theory (which, in fact, deals only with waves).

The facts could be stated in the language of photons as follows. A photon is detected, and we can state that this, with absolute certainty, is something other than a photon in a single beam because its behavior is determined by the physical properties of *both* beams. When the "identity" of the photon is checked by a suitable measurement (in the sense of belonging to one of the two beams, which have different propagation directions), we find that the interference pattern is destroyed. We could use also the following vague formulation: the fundamental uncertainty relating to which of the two beams the photon came from is an essential element of the interference process when we are dealing with independent photons. Similarly, the interference of a photon with itself is associated with a fundamental uncertainty about the path that the (one!) photon took.

Finally, let us not ignore the fact that additional confusion in the discussion about interference between independent photons was introduced by an often cited statement by Dirac (1958). Dirac not only asserted the interference of a photon with itself, but also declared that it is the only possible kind of interference. (His formulation was: each photon interferes only with itself; interference between different photons never occurs.) It follows from the context, however, that Dirac had a conventional interference experiment in mind.

Some researchers found it difficult to ignore this apodictic statement of such an authority as Dirac. There were attempts to "save" Dirac's statement by postulating that the emission from the two lasers cannot be guaranteed to be, in reality, independent; rather that each of the photons is generated *simultaneously* in both lasers and so interferes just with itself. The inadequacy of such an approach can be easily proven with the help of a delay line, for example, which can be used in such a way that two photons generated at different times interfere.

7.5 Which way?

We explained in the preceding section that experimental conditions under which single photon interference phenomena can be observed are quantum mechanically characterized by the impossibility of predicting anything about the photon paths. But why is this impossible? Let us first analyze the self-interference of the photon. In the Young double slit experiment we could use as the light source an excited atom and let the photon interfere, and *afterwards* we could "calmly" measure the recoil of the atom caused by the emission. The propagation direction of the photon is, due to momentum conservation, opposite to the recoil, and in this way we could find out after the event which of the slits the photon "really" passed through. The weak point of this argument is that Heisenberg's uncertainty relation for position and momentum of the atom is not taken into consideration, as was shown by Pauli (1933). Let us go into some detail.

To be able to observe interference the atom must be well localized. We analyze classically how the interference pattern will change when a point-like light source is shifted parallel to the interference screen by a distance δx (Fig. 7.5(a)). We will consider for simplicity a screen with two almost point-like holes, and we will assume further that the holes, the emitter and the point of observation lie in the same plane. Finally, we will assume for convenience that the emitter is positioned symmetrically between the two holes at position E_1. The intensity at the point of observation is determined by the difference between the two optical paths from the emitter through the holes H_1 and H_2, respectively, to the observation point. By shifting the emitter to position E_2 we change the two optical paths s_1 and s_2 from the emitter to the corresponding holes, and their difference (which is easily calculated) increases from zero to the value

$$s_2 - s_1 \approx 2\frac{d}{\sqrt{l^2 + d^2}}\delta x, \qquad (7.9)$$

where we have assumed $|\delta x| \ll d$; d is half the distance between the holes and l is the distance between the emitter and the interference screen. From Fig. 7.5(a) follows the relation $d/\sqrt{l^2 + d^2} = \sin \alpha$. Finally, dividing Equation (7.9) by the wavelength, λ, we obtain the following expression for the path difference change at the point of observation:

$$\delta g = \frac{s_2 - s_1}{\lambda} = \frac{2}{\lambda} \sin \alpha \, \delta x. \tag{7.10}$$

Repositioning the emitter has a minor influence on the position of the interference pattern only when δg is small compared to unity. We thus find that the atom must be localized with an accuracy

$$\Delta x \ll \frac{\lambda}{2 \sin \alpha} \tag{7.11}$$

(in the x direction) to avoid the washing out of the interference pattern.

Let us now discuss the experimental conditions for the measurement of the atomic recoil. The desired information about the direction of the emitted photon is delivered by the x component of the photon momentum (Fig. 7.5(b)). Due to momentum conservation, the atomic momentum change is equal in its absolute value to the photon momentum but has opposite sign. The photon momentum is given by $h\nu/c$ (see Section 6.9), and from Fig. 7.5(b) we find the momentum change of the atom in x direction to be

$$\delta p_x^{(2,1)} = \pm \frac{h\nu}{c} \sin \alpha, \tag{7.12}$$

where the plus sign applies to the lower and the minus to the upper light path.

The difference of the two momenta is

$$\delta p_x^{(2)} - \delta p_x^{(1)} = 2 \frac{h\nu}{c} \sin \alpha. \tag{7.13}$$

To distinguish between the two optical paths, we must measure the atomic momentum change δp_x with a greater accuracy than that given by Equation (7.13). We can imagine that we measure the atomic recoil with an arbitrary precision; however, what we are really interested in is the *change* in atomic momentum due to spontaneous emission. This means that our precision condition also concerns the atomic momentum *before* the emission: its uncertainty – we think again quantum mechanically – must be subjected to the constraint

$$\Delta p_x \ll 2 \frac{h}{\lambda} \sin \alpha \tag{7.14}$$

(and we assume that the mean value vanishes to get a static interference pattern). When we observe interference and also wish to measure atomic recoil, we run into a conflict with Heisenberg's uncertainty relation. Multiplying the two requirements

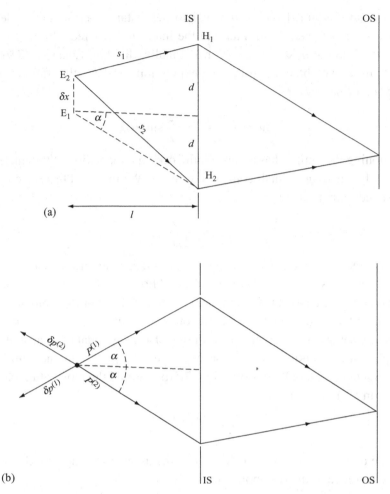

Fig. 7.5. Young's interference experiment. (a) Displacement of a point-like light source by δx. H_1 and H_2 = pinholes in the interference screen (IS); OS = observation screen. (b) Photon momenta $P^{(1)}$, $P^{(2)}$ and the corresponding atomic momentum changes $\delta p^{(1)}$, $\delta p^{(2)}$.

given in Equations (7.11) and (7.14), we obtain the inequality

$$\Delta x \, \Delta p \ll h, \tag{7.15}$$

which is in strict contradiction to Heisenberg's uncertainty relation

$$\Delta x \, \Delta p \geq \frac{h}{4\pi}. \tag{7.16}$$

The conclusion is that we can observe either interference or the emission direction thus determining the trajectory of the photon, as was claimed at the beginning.

Along similar lines, we can refute another objection which – among other penetrating attempts of Einstein to disprove quantum mechanics – was the subject of the famous Bohr–Einstein debate in 1927 at the Fifth Solvay Conference. Einstein's argument was based on the assumption that it is possible "in principle" to measure the recoil the interference screen suffers when the photon is passing because the photon changes its direction of propagation (see Fig. 7.5). To be able to ascertain from such a measurement the trajectory of the photon, we require, as in the previous case of recoil measurement, the initial momentum of the screen to be sufficiently well defined. According to Heisenberg's uncertainty relation, the position of the screen, and with it also the position of the two holes, is then not defined exactly, and a simple calculation reveals that it is uncertain to such an extent that the interference pattern is completely washed out.

While the experiments up to now have been only Gedanken experiments, recently experimentalists have become interested in the subject. A beautiful experiment was recently performed by American scientists (Zou, Wang and Mandel, 1991), who analyzed interference of light from different sources. The sources were non-linear crystals, each pumped by a strong coherent pump wave (laser radiation) exciting in them parametric fluorescence (discussed in Section 6.7); see Fig. 7.6. In this process, both a signal wave and an idler wave are generated. First, the two signal waves were made to interfere. This is by itself a surprising result. The observation of the interference pattern – resulting from many individual

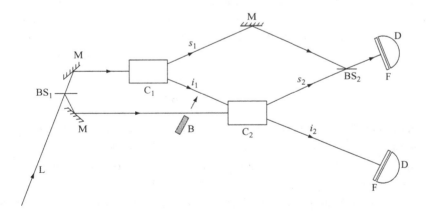

Fig. 7.6. Interference between two signal waves s_1 and s_2 generated through parametric fluorescence in two crystals C_1 and C_2. The interference disappears when the path of the idler, i_1, is blocked. i_2 is the second idler wave. L = laser beam; BS_1, BS_2 = beamsplitters; M = mirror; F = frequency filter; B = blocker; D = detector.

events – indeed required certain precautions to be taken: both crystals had to be pumped with the same laser radiation, and the geometry had to be set so that the idler waves coincided.

Let us start with the discussion of the first condition: that the crystals had to be pumped with the same radiation appears to be absolutely necessary because otherwise the emission processes in the crystals would be completely independent and no interference could be observed. One might also ask whether the phase correlations generated between the pump waves by the first beamsplitter are sufficient to generate the phase correlations between the signal photons that are needed for interference. The process of parametric fluorescence is definitely a spontaneous process! Indeed, the second condition, that the paths of the idler waves coincide, is necessary to guarantee the appearance of interference.

Let us look at the problem from a classical viewpoint. The crucial point is that, in the parametric process, the phases of the participating waves are related to each other in a certain way. Within the classical description we have to imagine that the crystal is illuminated by an intense pump wave together with a weak idler wave (or a weak signal wave) with a random phase. In contrast to quantum mechanics, where vacuum fluctuations of the signal and the idler wave are enough to "initiate" parametric fluorescence, we need in the classical theory a real wave (whose intensity can, in principle, be arbitrarily small) as a kind of nucleus, because the real wave must act jointly with the pump wave to generate in the medium a non-linear polarization oscillating at the signal wave frequency and hence acting as a source for the signal wave. (With the formation of a signal wave the idler wave is also amplified.) The phase relation in question comes about in the following way. Let us denote by φ_p and φ_i the pump and idler phases, respectively. The polarization induced by these two waves oscillates with the phase difference $\varphi_p - \varphi_i$, and because it is the source term for the emitted signal wave its phase is transferred to that wave (apart from a $\pi/2$ jump). The phase of the signal is hence given by

$$\varphi_s = \varphi_p - \varphi_i - \frac{\pi}{2}. \tag{7.17}$$

This relation applies to the processes taking place in both crystals. Assuming now that the propagation directions of both idler waves coincide, we deal with just a *single idler wave*. The statistical nature of its phase does not disturb the formation of a well defined phase relation between the signal waves, which can hence be made to interfere. Writing Equation (7.17) for the crystal C_1 at a time t_1 we see this quite clearly. The instantaneous phases propagate with the speed of light, hence the idler wave phases at C_1 and C_2, respectively, obey the simple relation $\varphi_i^{(2)}(t) = \varphi_i^{(1)}(t - \tau_{12})$, where τ_{12} is the propagation time taken by the light to travel from C_1 to C_2. In an analogous way, we can count back the pump wave phases at the crystal locations to the pump wave phase before the beamsplitter; we may write $\varphi_p^{(\mu)}(t) = \varphi_p^{(0)}(t - \tau_{0\mu})$, where $\tau_{0\mu}$ is the propagation time[6] between the beamsplitter and the crystal $C_\mu (\mu = 1, 2)$. According to Equation (7.17), we can write

$$\varphi_s^{(1)}(t_1) = \varphi_p^{(0)}(t_1 - \tau_{01}) - \varphi_i^{(1)}(t_1) - \frac{\pi}{2}, \tag{7.18}$$

$$\varphi_s^{(2)}(t_2) = \varphi_p^{(0)}(t_2 - \tau_{02}) - \varphi_i^{(1)}(t_2 - \tau_{12}) - \frac{\pi}{2}, \tag{7.19}$$

from which the signal wave phase difference follows as

$$\varphi_s^{(2)}(t_2) - \varphi_s^{(1)}(t_1) = \varphi_p^{(0)}(t_2 - \tau_{02}) - \varphi_p^{(0)}(t_1 - \tau_{01}) - \left[\varphi_i^{(1)}(t_2 - \tau_{12}) - \varphi_i^{(1)}(t_1)\right]. \tag{7.20}$$

As could have been expected, the correlation between the signal waves is most pronounced when the two times differ by exactly the propagation time between C_1 and C_2 ($t_2 = t_1 + \tau_{12}$). As in any conventional interference experiment the coherence length offers additional freedom for the interference observations. (The coherence length of the pump wave is much larger than that of the signal and idler waves, so we do not need to worry about it.) In the experiment, the expected periodic intensity dependence on the path difference of the two signal waves was indeed observed. (This was achieved by repositioning BS_2.)

It is without doubt that the correct way of describing this experiment must use the language of quantum mechanics. How do we discuss the problem quantum mechanically? First of all, we describe the pump wave classically, since it is an intense laser wave, and we represent it by an amplitude stabilized plane wave. Because the coherence length of laser radiation is large, another idealization is allowed, namely the phases of the two pump waves can be considered to be the same at the positions of the two crystals and, in addition, constant in time. The wave function of the radiation field emitted by each of the crystals is, in the lowest order of perturbation expansion (Ou, Wang and Mandel, 1989),

$$|\psi\rangle = |0\rangle_s|0\rangle_i + \beta A|1\rangle_s|1\rangle_i, \tag{7.21}$$

where β is a (positive) constant proportional to the non-linear susceptibility of the crystal and A is the complex pump wave amplitude. By doing this we have idealized the signal and idler waves as single-mode fields; this is justified by the fact that in the experiment a frequency filter was placed in front of the signal detector, making the signal waves quasimonochromatic. Due to the energy conservation for parametric processes expressed by Equation (6.8) and the sharp frequency of

---

[6] Any possible phase jump caused by beamsplitting is included in the propagation time.

the pump wave, the same applies to the idler waves. Equation (7.21) indicates that the information about the pump wave phase is "conserved" also in the quantum mechanical wave function.

Since in the discussed interference experiment the two idler modes coincide ($i_1 = i_2 = i$), the total wave function $|\psi_{tot}\rangle$ is not simply a product of two wave functions of the type (7.21), but takes the following form (under the assumption made above about the phase of the pump wave):

$$|\psi_{tot}\rangle = |0\rangle_{s_1}|0\rangle_{s_2}|0\rangle_i + \beta A(|1\rangle_{s_1}|0\rangle_{s_2}|1\rangle_i + |0\rangle_{s_1}|1\rangle_{s_2}|1\rangle_i). \qquad (7.22)$$

The vacuum term in the above equation is irrelevant for the photoelectric detection, as has already been mentioned. Only the second term is of importance for the description of the experiment. The common factor $|1\rangle_i$ of the two sum terms in the bracket is of no importance for the measurement on the signal waves, and so we come to the conclusion that the superposition

$$|\psi\rangle = |1\rangle_{s_1}|0\rangle_{s_2} + |0\rangle_{s_1}|1\rangle_{s_2} \qquad (7.23)$$

is responsible for the appearance of interference. However, this is exactly the same form of wave function that we encountered in the case of self-interference of a photon (see Equation (7.6)) and in the case of the interference of independent photons (see Equation (7.7)), which implies that the interference principle is always the same: interference is possible due to the fundamental impossibility in finding out which way the detected photon went, i.e. which source emitted it. In contrast, when such information can be obtained, the interference pattern will disappear. The experiment can illustrate this in an impressive way. The "birthplace" of the detected photon (C_1 or C_2) can be determined indirectly by performing a co-incident measurement on the corresponding idler photon. Observing, for example, that the idler photon has emerged from C_1 implies that the same must apply to the signal photon. To be able to make such a measurement, we have to modify the experimental setup shown in Fig. 7.6. Either we have to place a detector into the idler path between the two crystals (assuming the detector to be ideal, we can conclude from its non-response that the idler was emitted from C_2 and hence the second detector is unnecessary), or we misalign the crystals so that the idler paths become separated, enabling measurements by two separate detectors. In fact, the first setup does not even require a readout of the detector signal, it is enough to block the idler path by inserting an absorber. In the second case, the positioning of detectors is not required; even the insertion of blockers into the optical paths is unnecessary. The mere "threat" of performing such a measurement at any time is sufficient to make the interference disappear.

This is easily seen theoretically. From the classical viewpoint, in both cases the phase correlation between both idler waves is destroyed – we have to deal with two idler waves, instead of a single one, with randomly fluctuating phases, which, according to Equation (7.20), implies a similar behavior in the phase difference of the two signal waves. In the quantum mechanical description we have to work with two different idler modes, and Equation (7.22) is replaced by

$$|\psi_{tot}\rangle = |0\rangle_{s_1}|0\rangle_{s_2}|0\rangle_{i_1}|0\rangle_{i_2} + \beta A(|1\rangle_{s_1}|0\rangle_{s_2}|1\rangle_{i_1}|0\rangle_{i_2} + |0\rangle_{s_1}|1\rangle_{s_2}|0\rangle_{i_1}|1\rangle_{i_2}),$$
(7.24)

from which it is easily seen that the detector signal is no longer described by the superposition state in Equation (7.23) but by the corresponding mixture of states. It does not contain any phase information, and thus indicates that no interference takes place.

In the experiment by Zou *et al.* (1991), the interference pattern was destroyed by blocking the idler path. These researchers proceeded in fact in a more clever way. Instead of blocking the idler wave emitted from C_1 completely, they attenuated it gradually and observed a proportional decrease in visibility of the interference pattern.

The particular appeal of this experiment is that the destruction of the interference pattern is accomplished by affecting the idler waves only, leaving the signal waves undisturbed.

Finally, the experiment offers a good opportunity to introduce the concept of a "quantum eraser" (up to now it has remained as a Gedanken experiment; see Kwiat, Steinberg and Chiao (1994). The basic idea is as follows. The interference is first destroyed by obtaining information about the path of a self-interfering photon. When we "erase" this information, the interference pattern appears again. In our particular case we can extract the information about the path of the signal photon by inserting a $\lambda/2$ plate between the two crystals into the path of the idler wave i_1 (for aligned crystals). The polarization of the idler wave emitted by the first crystal is rotated by $90°$, and this wave can therefore be easily distinguished from the idler wave i_2 emitted by the second crystal. The determination of the polarization is realized with a polarizing prism (with detectors at each output) oriented such that its transmission directions coincide with the polarization directions of the two waves. The interference pattern which was present before thus disappears. The reason for this is clear: the two idler waves are statistically independent.

How do we eliminate ("erase") the which-way information? It is surprisingly simple: we just rotate the polarizing prism by $45°$. In each output the projections of the electric field strengths of the two idler waves onto the respective transmission directions are superposed, and we cannot conclude anything about the polarization state of the incident field and hence its place of origin. When we rotate the polarizing prism nothing happens to the signal photon, which was emitted a long time before, and probably does not exist anymore, having been absorbed by the detector. Hence we do not see a "revival" of the interference pattern. However, we can observe an interference pattern when *conditional* measurements are performed. We mean by this the following. We select those clicks made by the signal detector for which also, for example, the detector in the first output of the polarizing prism responded simultaneously. These clicks form an interference pattern, but only half of the signal photons contribute to its build up. The other half are recorded through coincidence measurements with the detector in the other output of the polarizing prism. The interference pattern is, in this case, shifted by half a fringe period from the one obtained previously. The superposition of the two patterns leads to its destruction – a maximum always meets a minimum. This must be so, since then the readings from the detectors – and hence the detectors themselves including the polarizing prism – can be dispensed with.

How can the selection carried out on the idler wave influence the signal wave? There cannot be an action back onto the emission process which has already ended! The key to an understanding of this problem is the correlation between the phase relation between the two idler waves and the phase relation between the signal waves. (See Equation (7.19), where $\varphi_i^{(1)}(t_2 - \tau_{12})$ should be replaced by the original value $\varphi_i^{(2)}(t_2)$.) The phase relation between the idler waves decides which output port of the polarization prism is taken by the idler photon. According to classical optics, for the special value 0 of the phase difference all the incident energy is concentrated in one of the outputs, while for a phase difference of π all the energy is concentrated in the other output. Hence, a selection of the relative idler phase is connected with the measurement of the idler photon, and because this phase was already fixed (randomly) in the emission process, we can make in this way a post-selection of the relative signal phase. This makes the appearance of interference at least qualitatively understandable and a change of the relative phase by π just leads to the shift of the interference pattern by half a fringe period, as predicted by quantum mechanics.

Finally, let us mention that atomic optics, which has undergone significant experimental progress in recent years, offers a very elegant example of the incompatibility between interference and which-way information. Due to the particle–wave dualism, material particles, such as electrons, neutrons or atoms, are also able to interfere with themselves. It is indeed possible to set up diffraction and interference arrangements for atomic beams based on optical models (Sigel, Adams and Mlynek, 1992). The place of light wavelength is taken by the De Broglie wavelength $\Lambda = h/p$, where h is Planck's constant and p is the particle momentum. Monochromaticity now means well defined velocity. The corresponding condition can be satisfied with high accuracy using the radiation pressure of an intense laser wave, which makes it possible, depending on the frequency detuning, to accelerate or slow down atoms. Atoms offer the great advantage of having an internal structure that allows, in particular, for selective excitation leading to spontaneous emission. The result is that the ability to interfere is either lost, or at least reduced (Sleator *et al.*, 1992). Indeed, a (possible) measurement on the emitted photons using a microscope would enable us to localize the emission center, i.e. the atomic center of mass (with precision limited by the wavelength of the emitted light). When the error of the position measurement is smaller than the distance between the interfering beams, we know the trajectory the atom had taken. Also in this case it is not necessary to perform the measurement. We can use the following argument (Paul, 1996). The spontaneous emission causes, through the recoil of the atom, a mechanical effect on the atomic center of mass motion. Let us consider for the moment only those events where the photon is emitted in a certain direction (the environment acts as a detector); then the atomic propagation direction becomes tilted (compared with the original horizontal direction). This causes an additional phase difference between the partial beams (let us consider, for instance, the interference on a double slit), resulting in a shift of the interference pattern. The total (final) interference pattern is composed from individual patterns shifted in different ways (corresponding to different emission directions of the photon), leading in every case to a deterioration of its visibility and, in the worst case, to its disappearance.

In contrast to the aforementioned optical experiment, the interfering particle suffers a massive disturbance so that the deterioration in its ability to interfere is not surprising. However, the atomic center of mass motion can be influenced in a much more subtle way, namely through microwave excitation of hyperfine levels. The transferred momentum is too small to influence noticeably the center of mass motion. Nevertheless, a clever manipulation through microwave pulses leads to the destruction of interference (Dürr, Nonn and Rempe, 1998). The information regarding which of two possible paths the atom had taken was stored in hyperfine structure states. The interference destruction was ultimately based on the fact that an atom, realized by a standing intense laser field of an appropriately chosen frequency, reflected from a beamsplitter and suffered a π phase shift exactly when it was in its lower hyperfine level.

7.6 Intensity correlations

The experimental technique described in Section 7.4 allows for direct observation of interference between strongly attenuated laser beams. There is, however, also an indirect method which we would like to describe in more detail. The basic idea behind the method is that two counters separated by a certain distance indicate the presence of an interference pattern, even when the pattern itself is running back and forth. First, let us assume that the detectors are positioned in such a way that the second counter is shifted – along the direction orthogonal to the (expected) interference fringes denoted as the z direction – with respect to the first one by a whole fringe separation Λ (or an integer multiple of it); then the counting rates will show strong correlations. This is not surprising, as the intensity incident onto each counter is the same, which also implies (for each moment) identical response probabilities. (It is as if both counters were at the same position.) On the other hand, when the separation of the counters, $\Delta z = z_2 - z_1$, is an odd multiple of half of the interference fringe separation, we will observe strong "anticorrelations." When the response probability of one of the counters becomes large, it is always small for the other one. The reason is clear: in such a case, one of the counters is near to an interference maximum while the other is near to an interference minimum.

A measure of the strength of the discussed anticorrelations is found experimentally in the following way. In a time interval of a prescribed length, we record the number of photons n_1 and n_2 detected by the first and the second counter, respectively; after repeating the experiment many times we obtain a data sequence with varying values of n_1 and n_2. From these data the so-called correlation coefficient k is determined, which is defined as follows:

$$k = \frac{\overline{\Delta n_1 \Delta n_2}}{\sqrt{\overline{\Delta n_1^2}}\sqrt{\overline{\Delta n_2^2}}}. \tag{7.25}$$

The bar stands for the average over the measured data, and $\Delta n_j \equiv n_j - \overline{n_j}$ ($j = 1, 2$) is the deviation of the individual measured value n_j from the mean value $\overline{n_j}$. The variance $\overline{\Delta n_j^2}$ is, as usual, defined as $\overline{\Delta n_j^2} = \overline{(n_j - \overline{n_j})^2}$.

The parameter k takes its maximum value for $\Delta z = n\Lambda$ ($n = 1, 2, \ldots$) and for $\Delta z = (n + \frac{1}{2})\Lambda$ it takes its minimum value. The latter value of k is, in contrast to the first, negative (hence the term anticorrelation). That this is indeed the case shows a simple classical analysis. We will assume that within the measurement interval the two interfering beams can be described by plane waves with well defined amplitudes and phases. To simplify the problem further, we will assume that both amplitudes are the same and constant and that only the phases change (uncontrollably) from one measurement to the other. The response probability of the photodetector is, in the classical description, proportional to the intensity, and we can use intensities instead of photon numbers. According to Equation (3.14), the intensities at positions z_1 and z_2 of the two detectors are (for $\overline{I_1} = \overline{I_2} = I_0$),

$$I_1 = 2I_0[1 + \cos(2\pi z_1/\Lambda + \Delta\varphi)], \tag{7.26}$$

$$I_2 = 2I_0[1 + \cos(2\pi z_2/\Lambda + \Delta\varphi)], \tag{7.27}$$

where we have replaced the z component of $\Delta \mathbf{k}$ by $2\pi/\Lambda$. For the mean value of $I_1 I_2$ with respect to $\Delta\varphi$, we find

$$\overline{I_1 I_2} = 4I_0^2 \left(1 + \frac{1}{2}\cos[2\pi(z_1 - z_2)/\Lambda]\right). \tag{7.28}$$

The variable $\overline{\Delta I_1 \Delta I_2}$ can be written as

$$\overline{\Delta I_1 \Delta I_2} \equiv \overline{(I_1 - \overline{I_1})(I_2 - \overline{I_2})} = \overline{I_1 I_2} - \overline{I_1}\,\overline{I_2}. \tag{7.29}$$

Inserting Equation (7.28) for $\overline{I_1 I_2}$ and the value $2I_0$ for $\overline{I_1}$ and $\overline{I_2}$ following from Equations (7.26) and (7.27), we obtain the expression

$$\overline{\Delta I_1 \Delta I_2} = 2I_0^2 \cos[2\pi(z_1 - z_2)/\Lambda], \tag{7.30}$$

confirming all the above statements about k.

Measurements of this kind were realized by Pfleegor and Mandel (1967, 1968) using two independent, strongly attenuated laser beams. They solved the problem of simultaneous photon counting at different places in a very elegant way by using a set of glass plates which were cut and arranged such that light incident on the first, third, fifth, etc. plate was directed towards the first detector, and light incident on the second, fourth, sixth, etc. plate was directed towards the second detector. To detect the desired anticorrelations, they had to measure n_1 and n_2 only within such time intervals for which the interference pattern changed very little. Information about the speed of drift of the interference pattern is naturally given by the frequency difference between the two interfering beams (see footnote 3 in Section 7.4). The procedure then was as follows. The unattenuated laser beams generated a beat signal in a photomultiplier, and this was used to control an electronic shutter; it was opened for 20 μs only when the beat frequency dropped below 50 kHz. During such a time interval each detector registered, on average, about five photons. The anticorrelation effect was indeed found in accordance with the theory; in particular, it turned out that the effect was maximal under the condition that the thickness of the plates was equal to half the fringe separation. In this way another, though indirect, proof of the interference between independent photons was given.

Similar conditions are present when the lasers are replaced by localized, excited atoms, each spontaneously emitting a photon. Certainly, both emission processes are mutually independent; in particular, there is no phase relation between the emitted waves. Under such circumstances anticorrelations of the intensity should be observable when adjusting the distance between the counters Δz to half the fringe separation Λ (with respect to the fictitious interference pattern generated by classical emitters radiating with a fixed phase difference). The measurement would proceed in such a way that coincidences, i.e. those events when both detectors simultaneously respond, are counted.

The principle of the effect, from a classical point of view, is easy to understand: when we identify the photons with classical waves with random phase values, then for the ensemble average – obtained from many repetitions of the experiment – the above Equation (7.28) applies. This implies that the coincidence rate (the number of registered coincidences per second) will decrease considerably when the distance between the two detectors changes from Λ to $\Lambda/2$.

Surprisingly, the effect becomes even more pronounced in the quantum mechanical description. The quantum mechanical treatment (Mandel, 1983; Paul, 1986) results in a modification of Equation (7.28), the factor of one-half in front of the cosine being replaced by unity (and the prefactor of four by two). While the classical theory predicts a drop of the coincidence rate to half of its mean value, according to quantum theory the coincidence rate completely vanishes for $\Delta z = (n + \frac{1}{2})\Lambda$ $(n = 0, 1, 2, \ldots)$. It is impossible to find two photons at a distance of $\Lambda/2$ (orthogonal to the fictitious interference fringes). This is indeed a specifically quantum mechanical effect, evading classical understanding. The surprising quantum mechanical result originates from the correct description of the physically obvious fact that "When two photons have been detected, *both* atoms had to deliver their energy because an atom can emit only *one* photon," while the classical description cannot rule out that the counts are due to the same atom. From this point of view, it is essential that we deal with *exactly two* atoms. The quantum mechanical description indeed goes over into the classical one for light sources consisting of many excited atoms. The discussed non-classical effect disappears also for the case when the numbers of atoms in both sources fluctuate according to a Poisson distribution. Then the mean atom number may be made arbitrarily small.

The described behavior of two spontaneously emitted photons must be completely incomprehensible within a naive photon picture. Imagining a photon as a bullet emitted in a certain (random) direction, we must admit that there should obviously be "communication" between the photons that no "mishap" takes place and the photons hit two positions $\Lambda/2$ apart, so breaking the rules of quantum mechanics. The described experiment does not allow us to determine the emission directions because we cannot know, in principle, from which atoms the detected photons have come. It is again the ignorance about the path – in this case of the two photons – that makes interference possible (in the sense of the aforementioned correlations). The correct description of the experiment requires a wave representation: the intensity correlations result from the superposition of the waves emitted by the atoms. From the total field thus formed, each of the detectors extracts an energy amount $h\nu$ (otherwise the event is not counted), and the question regarding the origin of the energy does not make sense.

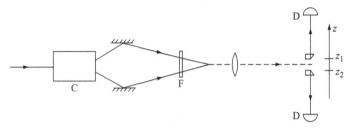

Fig. 7.7. Measurement of intensity correlations on two photons simultaneously generated by parametric fluorescence. C = non-linear crystal; F = frequency filter; D = detector.

In practice, there is not the slightest chance that the experiment would be possible. The emission takes place in the form of dipole waves, and because of this the probability of both detectors responding is extremely small. This difficulty can be overcome by using directed emission. As described, parametric fluorescence is a possible option at our disposal. An experiment of this kind was successfully carried out by Ghosh and Mandel (1987). The observation was made on two photons (the signal and idler photons) generated simultaneously by an incident strong ultraviolet laser beam in a non-linear crystal, which the photons left along slightly different directions (Fig. 7.7). They were made to interfere – in the sense of an intensity correlation – by two mirrors. The "interference pattern" was magnified using a lens to make the detection easier. Two movable optical glass plates defined the observation points z_1 and z_2. The incident light was directed to each detector by the plates. The coincidence was then determined electronically. The coincidence rate was modulated in dependence on the relative detector distance $z_2 - z_1$ as expected. The recorded measurement data – only a few events per hour were detected – were in quantitative agreement with the quantum mechanical prediction when it was taken into account that the observation points are determined with a limited precision given by the thickness of the glass plates.

As mentioned above, the intensity correlations are formed by superposing the field strengths of two independently generated waves on the detector. Such a superposition can also be achieved with the help of an "optical mixer." This is simply a beamsplitter with both input ports being used (Fig. 7.8): one of the fields is sent into the first port, and the other is sent into the second. In particular, a signal photon and an idler photon can be mixed in this way. The outgoing light is incident on two separate detectors. According to quantum theory, this mixing leads to a surprising result: the photons form a pair again (see Section 15.4). They leave the beamsplitter through the same output port, and it is naturally left to them to decide which of the output channels they use in a single measurement. For coincidence detection this means, however, that we find *with certainty* no coincidences!

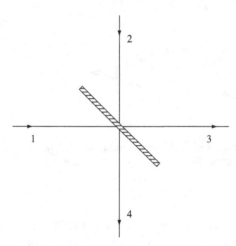

Fig. 7.8. Beamsplitter used as an "optical mixer." 1 and 2 are the incoming beams; 3 and 4 are the outgoing beams.

We have tacitly assumed an ideal alignment of the setup which guarantees that the photons indeed "meet" at the beamsplitter. In the opposite case, one photon takes no notice of the other and is transmitted or reflected with 50% probability. As a result, we have enough events where one of the photons hits the first detector and the other hits the second detector; we measure these coincidences. Their appearance indicates that the wave packets (pulses), as the photons appear to us in this kind of experiment, no longer overlap on the beamsplitter. We have a simple opportunity to measure the effective length of a wave packet: displacing the beamsplitter (Fig. 7.9) makes the lengths of the two light paths different, and the coincidence rate rises from its minimal value zero to a finite value and stays constant. The width at half maximum of the enveloping curve – as a function of the beamsplitter position – gives us a measure of the wave packet's length, i.e. its spatial extension. This experiment was in fact performed by the American scientists Hong, Ou and Mandel (1987). For the time spread of the photon they found a value of about 50 ps, which was determined essentially by the transmission width of the frequency filters positioned in front of the detectors (according to the general relation Equation (3.22) between the frequency width and the time spread of a wave packet). The technical measurement limit for the determination of the pulse length was in fact much lower (about one femtosecond). It is determined by the precision with which the displacement of the beamsplitter can be measured.

The point of this experiment is that it allows the measurement of extremely short times with "inert" counters. In fact, the integration time is required only to be shorter than the distance between successive pulses in order to ensure that only one photon pair is registered.

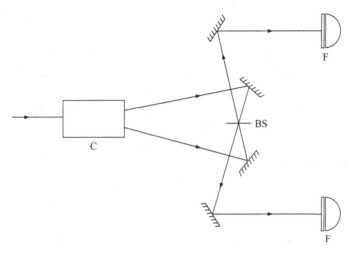

Fig. 7.9. Experimental setup for the measurement of the spatial extension of signal and idler photons. C = non-linear crystal; BS = beamsplitter; F = frequency filter.

7.7 Photon deformation

In Section 7.2 we explained in detail that single photons exhibit interference phenomena which are described correctly by the classical wave theory. According to this theory conventional interferometers work well even when single photons (in principle with arbitrarily large separation) are sent through them. However, as we explained in Section 3.4 for the example of the Fabry–Perot interferometer, the frequency narrowing goes hand in hand with a lengthening of the incident light pulse. Single photons are already subjected to this process. This means that the photon – considered as a wave – is deformable.

The change in form is not limited – recall the Fabry–Perot interferometer – to a stretching of the wave packet. Only a frequency section of the incident wave is transmitted; the remaining part is reflected. (We need not take into account events when the photon is absorbed in one of the silver layers because then it is "eliminated completely.") Even though both parts of the wave can become arbitrarily far apart as time passes, they must be considered *together* as a wave phenomenon associated with a single photon. Only after a measurement has taken place is the photon found in one or the other partial beam.

If photons – as waves – can become longer, there is no doubt that it must be also possible to make them shorter. This can be accomplished with a fast shutter which can cut parts of the wave off or out. An impressive experiment of this form was performed by Hauser, Neuwirth and Thesen (1974), who "chopped" γ-quanta using a very fast rotating wheel with absorbing spokes and observed that the resulting broadening of the spectral distribution was in excellent agreement with the prediction of classical wave theory (see Equation (3.22)).

We encountered a situation similar to spectral decomposition in the direction selection of photons. Let us consider a photon having the form of a spherical wave incident on a reflecting screen with an aperture. The corresponding wave is split into a reflected part and a transmitted part. Because the two parts can be made to interfere at any time, it would certainly be wrong to imagine the photon to be "in reality" only in one of the partial beams while the other is "empty." When we perform a measurement on one of the partial beams (with a photodetector), either we detect a photon or we do not. In the latter case – assuming ideal detection efficiency of the measurement apparatus – we can conclude that the photon is present in the other beam.

It is important to point out in this context that the measurement process in the quantum mechanical sense is not necessarily associated with sophisticated and complicated apparatus. In fact, for a photodetector, the elementary process of electron release from the atom should be viewed as the actual measurement act. More generally, any absorption process, provided it is irreversible (the absorbed energy is transferred to the surroundings and thus dissipated), can pass for a measurement process. For example, when, in the spectral decomposition of light with a Fabry–Perot interferometer, the reflected light is incident on an absorbing screen, a measurement takes place which determines whether the photon was reflected or not. This measurement "allowed survival of" only those photons which took the path through the interferometer and hence have a sharper frequency than at the beginning. Similarly, the propagation direction is selected, in the case of a photon, in the form of a spherical wave, incident on an absorbing screen with a small aperture (Renninger, 1960).

It might seem paradoxical that the physical properties of the outgoing photons were changed even though the screen obviously did not interact with them. However, it would be wrong to state that nothing happened under such conditions. In such a case – let us consider direction selection – the photon would have to possess a well defined propagation direction before it was incident on the screen. This is, however, in contradiction to the fact that we can perform interference experiments with the original photons, for example by following the example of Young, who illuminated a screen with two holes and observed the transmitted light on an observation screen.

The question of how the photon is affected by the screen when it passes the hole remains unanswered by quantum mechanics. It just "manages," with the aid of simple axioms, to predict exactly the *experimentally verifiable* effects of a measurement apparatus – an achievement coming close to a wonder in the face of the complexity of the measurement process! An essential role in this scenario is played by the "collapse of the wave function" describing the transition from the possible to the factual. To make it explicit using the example of the direction selection: the spherical wave "contains" all possible propagation directions, and the observation apparatus forces a "decision" regarding which of them becomes real. (Note that the position on the screen where the photon absorption took place can be determined in principle, and we can later – for a known position of the emission center – determine the propagation direction of the particular photon.) The manipulation possibilities described in Section 6.7 are, in the case of parametric fluorescence, simply astonishing. They are based on the fact that signal and idler photons are in an "entangled" quantum mechanical state. Through the measurement, for example on the corresponding idler photon, signal photons with desired properties such as spatial localization or sharp frequency can be selected.

There are no indications that quantum mechanics is just a "temporary solution" – in particular the real Einstein–Podolsky–Rosen experiments described in Section 11.4 let all hope fade for a refining of the quantum mechanical description by introducing "hidden variables" – and we have, with regret, to accept the fact that Nature is not willing to "disclose its secrets" on the microscopic level as it does on the macroscopic level.

8

Photon statistics

8.1 Measuring the diameter of stars

As mentioned several times already, the particle character of light is best illustrated by the photoelectric effect. This effect can be exploited in the detection of single photons by photocounting. The analysis of such counting data allows us, as will be discussed in detail in this chapter, to gain a deeper insight into the properties of electromagnetic fields. We can recognize the "fine structure" of the radiation field – in the form of fluctuation processes – which was hidden from us when using previous techniques relying only on the eye or a photographic plate, i.e. techniques limited to time averaged intensity measurements.

The credit for developing the basic technique for intensity fluctuation measurements goes to the British scientists R. Hanbury Brown and R. Q. Twiss, who became the fathers of a new optical discipline which investigates statistical laws valid for photocounting under various physical situations. When we talk of studies of "photon statistics" it is these investigations that we are referring to.

Interestingly enough, it was a practical need, namely the improvement in experimental possibilities of measuring the (apparent) diameters of fixed stars, that gave rise to the pioneering work by Hanbury Brown and Twiss. Because the topic is physically exciting, we will go into more detail.

It is well known that the angular diameters of fixed stars – observed from Earth – appear to be so small that the available telescopes are not able to resolve the stars spatially. The starlight generates a diffraction pattern in the focal plane of the telescope, the form of which is determined by the aperture of the telescope and has nothing to do with the real spatial extension of the star. A solution to this problem was required before any progress could be made. An idea by Fizeau led to a practical solution in the form of Michelson's "stellar interferometer."

In this apparatus, the starlight falls onto two mirrors, M_1 and M_2, a distance d apart, and the light is redirected by mirrors M_3 and M_4 into a telescope and focused

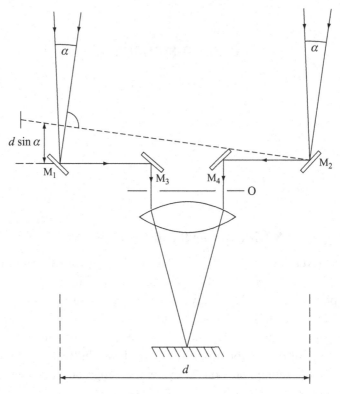

$d \sin \alpha$

Fig. 8.1. Michelson's stellar interferometer. O = diaphragm; M_1, M_2 = movable mirrors, M_3, M_4 = fixed deviating mirrors.

in its focal plane. In addition, filters are inserted into the optical path to guarantee that only light of a certain frequency is observed (see Fig. 8.1).

To understand how the stellar interferometer works, let us assume, at first, that the source is point-like. The large distance between the stars and the Earth means that the light beams incident on M_1 amd M_2 are essentially parallel. As in the Michelson interferometer, an interference pattern in the form of straight line, equidistant fringes appears in the focal plane of the telescope. The fringes are formed because M_1 and M_2 are orientated at an angle not exactly 45° with respect to the telescope axis. Due to this we are dealing with interference lines of equal thickness, similar to those to be observed on a wedge. Because we are dealing with a light source of finite spatial extension, we can imagine it as consisting of individual parts, each of which generates by itself an interference pattern of the described form. It is of importance that usually the interference patterns do not coincide precisely in their positions but are mutually slightly shifted. The reason for this is that two interfering beams emitted from one part L_1 of the light source have an additional path difference Δs compared with two other interfering beams coming from another part of the light source L_2 (imaged on the same point in the focal point of the telescope). As can be seen from Fig. 8.1, the path difference Δs is given by

$$\Delta s = d \sin \alpha \approx d\alpha, \tag{8.1}$$

where α is the (very small) angle between the two directions under which the light beams from points L_1 and L_2 are incident upon the surface of the Earth.

The superposition of the interference patterns related to different parts of the light source does not, according to Equation (8.1), play a significant role as long as the product of the mirror distance d and the maximum value α_0 of angle α, which obviously should be identified with the angular diameter of the star, is small compared with the wavelength λ of the starlight. Moving M_1 and M_2 further and further apart from one another causes the visibility of the interference pattern to become worse – due to an increase of the relative shifts between the individual patterns – until no interference fringes can be observed. As a rough estimate this happens when

$$d\alpha_0 \approx \lambda. \tag{8.2}$$

For this situation the interference patterns from the right and left edges of the star's surface coincide, whereas the patterns generated by light emitted from the other parts of the star's surface are shifted in all possible ways. Hence the superposition of the individual patterns results in a uniform distribution of brightness – the interference has disappeared.

We have already mentioned a possible way of measuring the star's diameter; i.e. increasing the mirror separation d in the Michelson stellar interferometer until no interference pattern is observable. Inserting the experimentally found critical value for d into Equation (8.2), we obtain the value of the star's angular diameter α_0. A refined version of this simplified approach allows the calculation of the visibility of the interference pattern as a function of the mirror separation d for a given brightness distribution of the star's surface (see Mandel and Wolf, 1965). In particular, it can be shown that Equation (8.2), for the case of a uniformly radiating circular disk, must be corrected as follows:

$$d\alpha_0 = 1.22\lambda. \tag{8.3}$$

The Michelson method proved to be very successful, and star diameters down to about 0.02 arc seconds have been determined (Michelson and Pease, 1921; Pease, 1931). An additional increase of the resolving power, requiring mirror separations of many meters, faces two practical obstacles.

First of all, the finite linewidth (ignored up to now in the analysis) of the observed starlight causes trouble: it decreases the visibility of the interference pattern when the light paths – going through M_1 and M_2, respectively – of the interfering beams uniting in the focal plane of the telescope do not have exactly the same length. Under such circumstances, a small change in the wavelength causes a shift in the corresponding interference pattern. To eliminate this disturbing effect for a typical bandwidth of $\Delta\lambda = 5\,$nm we would have to guarantee the path difference to be less than 0.01 mm. Such a condition imposed on the mechanical stability of the interferometer – and fulfilled during the duration of the observation usually also involving guiding the instrument – is absolutely impossible for the required large length of the interferometer arms.

Second, the observation of the interference pattern is hindered by so-called atmospheric scintillations. By this we mean the influence of atmospheric fluctuations, i.e. local air motion altering the air pressure, leading to changes of the air's index of refraction. These atmospheric disturbances, for large mirror separations, are statistically independent for the two light beams and cause fluctuations of the path difference, and therefore the position of the interference pattern changes in time in an uncontrollable way.

The difficulties described above were overcome by Hanbury Brown and Twiss, who extended their radioastronomical method developed a few years earlier to the optical domain. They measured correlations between intensities at the different positions rather than between electric field strengths (which is the essence of any interference experiment). In this way they freed the experiment once and for all from any disturbances dependent on phase fluctuations, simply because the phase no longer appeared in the measurements. The experiment was arranged in such a way that the outer planar mirrors in Michelson's setup were replaced by parabolic mirrors (Fig. 8.2) which focused the incident starlight onto separate photomultipliers (Hanbury Brown and Twiss, 1956b). The correlations between the photocurrents were measured such that they were each amplified in the narrow band and then multiplied. The measurement signal was the time averaged value of their product. The photocurrent follows the time fluctuations of the light incident on the photocathode (as explained in Section 5.2), and hence the signal is a measure of the time averaged value of the product of the light intensities at the positions of the detectors, and thus it reflects the intensity correlations in the radiation field. The photocurrents are combined using normal electric wires, which eliminates the need for mechanical stability of the interferometer setup.

We might ask whether it is really possible to recover the information contained in the visibility of interference fringes (and determined in the first place by phase relations) from the intensity correlations. In the following, we show that this is indeed possible.

Let us once again return to the Michelson measurement technique. In classical optics the ability to interfere is synonymous with coherence, and so we can say

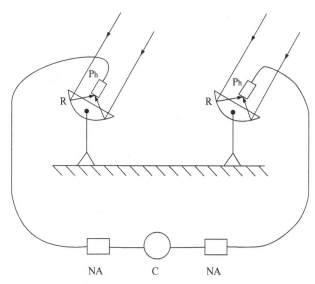

Fig. 8.2. Hanbury Brown–Twiss stellar interferometer (Ph = photomultiplier; C = correlator; R = reflector; NA = narrow-band amplifier). The product of the two photocurrents is taken and time averaged in the correlator.

that the Michelson stellar interferometer analyzes the spatial coherence of starlight (transverse to its direction of propagation). In particular, the transverse coherence length is measured, which is simply the critical mirror separation for which the interference vanishes. Spatial coherence primarily means that the (instantaneous) phase of the electric field strength changes only slightly within the coherence area. Since the stars are thermal radiators, the phase at a given position changes in an uncontrollable way, but the important point is that the electric field strength in the neighborhood – within the coherence area – follows precisely these phase fluctuations.

The thermal radiation field exhibits not only phase fluctuations but also strong amplitude fluctuations. The instantaneous amplitude also changes noticeably over finite distances only, and it is – fortunately for the Hanbury Brown–Twiss procedure – a characteristic feature of thermal radiation that the coherence area for which the phase is approximately constant coincides (apart from minute details) with the spatial domain for which the amplitude is also approximately constant. Observing the time evolution of the intensities at two different points P_1 and P_2 lying on a plane (almost) orthogonal to the propagation direction, we find that the instantaneous intensities, in most cases, coincide, provided the distance d between P_1 and P_2 is shorter than the transverse coherence length l_{trans}. When we experimentally find an intensity peak at P_1, then there is a large probability that we will find the same at P_2, and the same applies to an intensity "dip" (a value below the average intensity).

Roughly speaking, the following relation holds

$$\Delta I_1(t) \approx \Delta I_2(t) \quad (\text{for } d \leq l_{\text{trans}}), \tag{8.4}$$

where we have defined the deviation from the time averaged value $\overline{I}$ by $\Delta I(t) \equiv I(t) - \overline{I}$ and the subscripts refer to the positions P_1 and P_2. As already discussed, the correlator used by Hanbury Brown and Twiss ultimately registers the time averaged product of $I_1(t)I_2(t)$ which, using Equation (8.4), can be written as (note that $\overline{\Delta I} = 0$)

$$\overline{I_1(t)I_2(t)} = \overline{I}^2 + \overline{\Delta I_1^2} \quad (\text{for } d \leq l_{\text{trans}}). \tag{8.5}$$

When the distance between P_1 and P_2 exceeds the transverse coherence length l_{trans} the intensities at the two points fluctuate independently, and the relation

$$\overline{\Delta I_1(t)\Delta I_2(t)} = 0 \tag{8.6}$$

holds, which results in

$$\overline{I_1(t)I_2(t)} = \overline{I}^2 \quad (\text{for } d > l_{\text{trans}}). \tag{8.7}$$

The decrease in the intensity correlations (with increasing d) allows us to draw conclusions about the coherence length l_{trans}, and by inserting it into Equation (8.3) in place of d we can find the star's angular diameter. Hanbury Brown and Twiss (1956b) estimated the diameter of Sirius in exactly this way. The possibilities offered by the method were further exploited in the early 1960s when an extended observatory was installed at Narrabri, Australia (Hanbury Brown, 1964). The parabolic mirrors were mounted on railway bogies moving on circular tracks with a diameter of 188 m. This setup allowed the measurement of star diameters down to 0.0005 arc seconds.

The intensity correlations can also be detected using photo counters – and this brings us finally to photon statistics. The response probability of a photodetector is proportional to the instantaneous intensity on its sensitive surface, and so the number of photons $n(t; T)$ counted within a finite time interval $t - \frac{T}{2}$ to $t + \frac{T}{2}$ reflects the intensity of the field. (The integration time T must be chosen to be smaller than the coherence time of the field, otherwise the intensity fluctuations will be averaged out.) The intensity correlations are now determined in such a way that we form the product $n_1(t; T)n_2(t; T)$ of the registered photon numbers $n_1(t; T)$ and $n_2(t; T)$ and average over a longer set of measured data.

The same physical information can be obtained by counting coincidences; i.e. we register only those events when both counters respond at the same time. In fact, the probability for such a type of coincidence is proportional to the time average of $I_1(t)I_2(t)$. With the stellar interferometer technique of Hanbury Brown and Twiss an increased number of coincidences is observed when the distance between the two detectors is smaller than the transverse coherence length. When the distance between the detectors exceeds this critical value, the intensities at the two positions fluctuate independently and the coincidences are purely random (see Equation (8.7)). The result of the observations can be formulated as follows: for $d \leq l_{\text{trans}}$ more coincidences will be detected than from the random case.

There exists a correlation, for $d \leq l_{\text{trans}}$, between photons measured at the same time at different points: when a photon is observed at position P_1, the probability of detecting a photon at P_2 is larger than for the completely random case.

Such a fact is incomprehensible when we employ a naive photon concept that excludes wave properties of light. We can imagine that the atoms on the stellar surface radiate independently, thereby each emitting – in any elementary act of emission – a photon. However, if nothing else happens and the photons fly through space like little balls without affecting each other and then hit the Earth's surface, it would be impossible to understand how the described correlations could be formed. A single photon cannot "know" what else is happening in its surroundings. In reality, however, the statistical behavior of the photons is determined by the diameter size of the stars!

These facts can be understood only in the framework of the wave concept of light. A photodetector is influenced by the instantaneous intensity of the electromagnetic field residing on its sensitive surface. The intensity is determined by the electric field strength, which is given as a superposition of many elementary waves – in principle *all* the atoms on the star's surface contribute – and this is the deeper physical reason why the intensity correlations contain information about the spatial extension of the surface of the star.

In fact, the spatial coherence (in the transverse direction) measured in Michelson's stellar interferometer comes about by the same mechanism (Fig. 8.1): the electric field strengths at the mirror positions P_1 and P_2 can be correlated only because the radiation in both cases is coming from the *same* atoms. The elementary waves emitted by the atoms can have, as is indeed the case, phases and amplitudes that are fluctuating randomly and independently. These fluctuations (we assume that the distance between P_1 and P_2 does not exceed the transverse coherence length) cause the same effect at the two observation points, and due to this the fluctuations of the total electric field at the two points have "the same beat."

This illustrates once again that we cannot assign an individuality to the photons by ascribing to each of them a certain "place of birth" (a well defined, though unknown, atom). The conclusion drawn in Section 7.4 in connection with the interference between independent photons was very similar. We can state quite generally that the naive photon picture always fails when interference is involved.

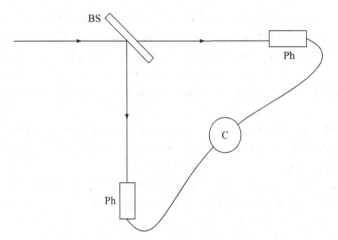

Fig. 8.3. Hanbury Brown–Twiss experimental setup for intensity correlation measurement. Ph = photomultiplier; BS = beamsplitter; C = correlator.

Finally, let us mention that Hanbury Brown and Twiss (1956a) tested the proposed measurement technique in the laboratory before they made their astronomical observations. As with the intended astronomical measurements, they analyzed the transverse spatial coherence of a thermal radiation field, namely that of a diaphragm placed before a mercury lamp. Naturally the coherence length is, under such circumstances, very small, and therefore the two receivers could not be positioned next to each other. Hanbury Brown and Twiss overcame this difficulty in an elegant way by using a beamsplitter (Fig. 8.3). As expected, they observed a decrease in the intensity correlations when one of the two photomultipliers was displaced sideways from the position corresponding to the mirror image of the other detector. The experiment provided clear proof of the existence of intensity correlations (for thermal radiation) and became a prototype for later photon statistical experimental setups. As Rebka and Pound, (1957) showed for the first time, such a setup is also well suited to the observation of temporal intensity correlations, and we devote the following section to this problem.

8.2 Photon bunching

The dependence of the spatial intensity correlations on the separation of the detectors discussed up to now is determined by geometric parameters, i.e. the size of the light source and its distance from the detectors. It can also be expected that time dependent correlations can appear which originate in the time fluctuations of the light intensity (at a fixed position). The mean time spread of an intensity maximum or minimum for thermal light is approximately equal to the time interval in which

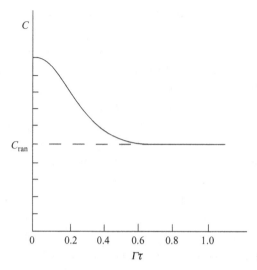

Fig. 8.4. Theoretical dependence of the coincidence counting rate C on the delay time τ for polarized thermal light with a Gaussian spectral profile $G(\nu) = \text{const.} \exp(-(\nu - \nu_0)^2/\Gamma^2)$. C_{ran} = random coincidence counting rate. After Mandel (1963).

the phase changes only slightly. This means that it coincides with the coherence time t_{coh}. (An analogous statement was made in the discussion of spatial intensity correlations.) Because t_{coh} is, in order of magnitude, given by the inverse of the linewidth $\Delta\nu$, the spectral properties of light come into play in the analysis of time dependent intensity correlations. The observation of time delayed coincidences at a fixed point requires the measurement of the time averaged value of $I(t)I(t + \tau)$ (τ is the delay time). The coincidence rate found for $\tau < t_{\text{coh}}$ is higher than that for $\tau > t_{\text{coh}}$ because $\overline{I^2(t)}$ is larger than $\overline{I(t)I(t + \tau)}$ for $\tau > t_{\text{coh}}$. (The relation is very similar to that for spatial intensity correlations. The considerations leading to Equations (8.5) and (8.7) can be applied directly to the present situation.) The coincidences are, for $\tau \gg t_{\text{coh}}$, purely random, and therefore an excess of coincidences compared with the random ones can be observed only if the delay time stays below the coherence time. For thermal radiation with a Gaussian spectral profile we expect theoretically a coincidence dependence of the type given in Fig. 8.4. Herewith, a new possibility for the measurement of linewidth of thermal radiation via time dependent intensity correlations appears. The experimental setup of Hanbury Brown and Twiss shown in Fig. 8.3 offers a possible experimental realization. Both detectors are now fixed – their positions are mirror images with respect to the beamsplitter – and the measurement is carried out in such a way that, from the whole ensemble of events detected by the counters, those are selected (electronically) for which the second detector responded τ seconds after the first one. (When using photomultipliers, one of the two photocurrents is delayed before multiplication in the correlator.)

As explained above, the temporal intensity correlations have a duration of about t_{coh} (see Fig. 8.4). The measurements require detectors with a response time shorter than t_{coh}. This implies that measurements of intensity correlations allow the determination of very small linewidths because in that case t_{coh} is very large. In contrast, conventional optical spectrometers, for example the Fabry–Perot interferometer, are well suited for the measurement of large linewidths and line distances. The new techniques based on photon statistical measurements and the traditional interferometric techniques fortuitously complement each other very well.

In fact, the spectral lines emitted by thermal radiators are so broad that the detection of intensity correlations is very difficult. However, there is a rather important application for the new technique, namely the analysis of laser light, scattered by moving centers. In contrast to the incident (monochromatic) light, the scattered light has the same character as thermal light. The reason for this is simple: the scattered light – which is very similar to thermal radiation – is formed from many partial waves with random phases emitted from individual centers. (There are fixed phase relations with respect to the incident laser radiation, which has a phase that is constant over its cross section and changes only slowly in time, but the irregular distribution of the centers causes a random distribution of scattered wave phases when observed from a randomly chosen fixed point.) The result is (at a fixed time) the formation of a "light ridge" in space. Due to the random motion of the scattering centers (for example resulting from Brownian motion) the intensity also fluctuates in time in the same way as it is known to do for thermal light. Because the scattered light has an extremely narrow bandwidth, it is almost an ideal object for photon statistical observations. For example, it is possible to determine diffusion coefficients of particles undergoing Brownian motion in liquids by measuring time dependent intensity correlations. A typical measurement curve is given in Fig. 8.5. (The exponential decrease for increasing delay time is caused by a Lorentzian line profile, in contrast to the Gaussian profile assumed in Fig. 8.4.) Similarly, the heat conductivity of liquids can be determined from measuring Rayleigh scattering of laser radiation caused by temperature fluctuations.

The situation illustrated in Fig. 8.5 can be visualized as a grouping – or "bunching" – of photons: the probability that two photons arrive at the same place shortly after one another (i.e. with a defined delay $\tau < t_{coh}$) is distinctly higher than the probability that they arrive with a larger time delay $\tau > t_{coh}$. We can also say that the photons tend to appear in pairs.

This phenomenon seems surprising from a naive photon picture. Let us again consider a thermal source: it seems that after the first atom has emitted a photon, at least one more atom hurries to emit another photon. In fact, "photon bunching"

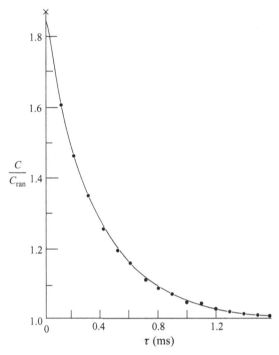

Fig. 8.5. Ratio of the coincidence counting rate C and the random coincidence counting rate C_{ran} as a function of the delay time τ measured on a suspension of bovine serum albumine (Foord *et al.*, 1969).

only reflects the strong intensity fluctuations (as do the spatial intensity correlations described in Section 8.1), and therefore it is the consequence of the interference between individual elementary waves emitted *completely independently* by different atoms.

Another photon statistical measurement method is based on the registration of the number $n(t; T)$ of photons during a fixed time interval of length T and than extracting from such a data sequence the probabilities that exactly 0, 1, 2, ... photons have been observed. From a theoretical point of view, as will be explained in detail below, we expect the photon numbers $n(t; T)$ for polarized thermal light to obey a Bose–Einstein distribution; i.e. the probability p_n (normalized to unity) of finding exactly n ($= 0, 1, 2, \ldots$) photons is given by

$$p_n = \frac{\bar{n}^n}{(\bar{n} + 1)^{n+1}}, \tag{8.8}$$

where $\bar{n}$ is the mean photon number. Interestingly enough, the distribution has its maximum for the value $n = 0$ (Fig. 8.6); i.e. the probability of not finding any photons is larger than the probability of finding any *given* finite number of photons.

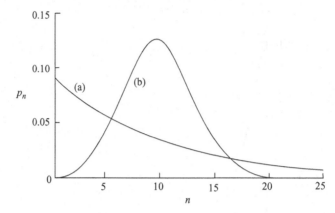

Fig. 8.6.   Bose–Einstein distribution (a) and Poisson distribution (b) for the average photon number $\bar{n} = 10$. p_n is the probability of finding n photons.

The variance of the photon number for the distribution function Equation (8.8) is

$$\Delta n^2 \equiv \langle n^2 \rangle - \langle n \rangle^2 = \langle n \rangle^2 + \langle n \rangle. \qquad (8.9)$$

This result tells us that the photon number fluctuates strongly. This is a characteristic of thermal light. Let us note at this point that there is a distinct difference between polarized and unpolarized light, with polarized light (which we have considered in this case) fluctuating much more strongly than unpolarized light.

Let us also comment on the experiment. First, we note that Equation (8.8) was derived within a single-mode approximation. This formula is simply the well known Boltzmann distribution $p(E_n)$ for the energy level occupation of a system in thermodynamic equilibrium. In our case, as explained in Section 4.2, the levels are equidistant – so the relation $E_n = nh\nu$ ($n = 0, 1, 2, \ldots$) holds – and hence the Boltzmann factor is

$$p(E_n) = \text{const.} \exp\{-nh\nu/k\Theta\}, \qquad (8.10)$$

where k is Boltzmann's constant and Θ is the temperature. Rewriting the relation for the mean photon number $\langle n \rangle$ (and properly normalizing) we indeed obtain Equation (8.8). Our considerations make it clear that the photon number is primarily related to the *mode volume*, which can be identified with the coherence volume V_{coh}. In any case, we must take care in experiments that the above summation over the individual events is limited to beam volumes smaller than V_{coh}. In particular, this implies that the integration time T must not exceed the coherence time. So, what happens for smaller T, i.e. when only a part of the coherence volume V_{coh} is under observation? Obviously, only a fraction of the photons present in the coherence volume V_{coh} are observed. However, it would be erroneous to think that this fraction is fixed. In this case randomness also comes into play, and the selection mechanism is of the same type as that for beamsplitting described in Section 7.1. (Instead of having transmitted and reflected photons, we divide them into observed and unobserved ones.) As mentioned at the end of Section 7.1, the process transforms thermal light back into thermal light. This means that Equation (8.8) also applies – with a correspondingly decreased photon number $\langle n \rangle$ – to the photons selected by the detector.

With respect to real measurements, we also have to take into consideration the fact that the detection efficiency of real detectors is considerably smaller than unity; i.e. an incident photon is detected only with a certain (fixed) probability. This makes it possible to model a less efficient detector by a combination of a perfect detector and an absorber placed in front of it. Because the absorption process is formally the same as the process of beamsplitting, the inefficiency of a photo-detector has the same effect as the previously analyzed decrease of the observation volume compared to the coherence volume.

We can thus conclude that even inefficient detectors do not change the character of thermal light. Equation (8.8) is also valid for the actually measured photons. However, we must not forget one point: the integration duration T must be smaller than the coherence time, i.e. the inverse of the linewidth. Like the observation of photon "bunching," the measurement of the photon distribution Equation (8.8) is possible only for thermal light with a very narrow linewidth.

This is the reason why in this case one works with scattered laser light. Arecchi, Giglio and Tartari (1967b) used as the scattering medium a suspension of polystyrene balls of different size in water and demonstrated that the photon number distribution $n(t; T)$ of the scattered light obeyed a Bose–Einstein distribution.

There is a rather simple way of generating narrow-band "pseudothermal" light for demonstration purposes: let quasimonochromatic laser light be reflected from a rotating ground glass plate (Martienssen and Spiller, 1964). The incident light is scattered from an irregular rough surface. The rate of the amplitude change obviously depends on the angular velocity of the plate and hence can be arbitrarily varied. The coherence time t_{coh} of the scattered light can be adjusted in such a way that photon statistical measurements can easily be carried out. In fact, the first successful experimental proof of the Bose–Einstein distribution of photons was given in this way (Arecchi, Berne and Burlamacchi, 1966).

It is very important to note that the previously described photon statistical properties of light are characteristic of radiation emitted by thermal sources (and sent through a polarizer) or of the aforementioned scattered radiation. In fact, it is possible, using a laser, to generate light with completely different properties. Before we go into details of this problem, let us comment on the influence of inefficient detectors on the outcome of coincidence measurements.

First, we have to know the dependence of the simultaneous response probability of two ideal detectors at the same position on the photon number n in the given mode volume or coherence volume.

Following classical arguments, we are tempted to say that the probability in question is proportional to the time averaged square of the (instantaneous) intensity, or – since the intensity and the photon number are related to each other through a constant factor – proportional to the time averaged value of n^2, which can be replaced, due to the ergodicity of the radiation field, by the ensemble average $\langle n^2 \rangle$.

In contrast, the quantum mechanical description leads to the result that, in the expression for the coincidence counting rate, the term $\langle n^2 \rangle$ must be replaced by $\langle n(n-1) \rangle$ (see Section 15.1). The quantum mechanical result is more trustworthy than the prediction based on classical considerations because it takes the energy conservation law into account. This can be seen most easily by considering the situation with exactly one photon in the coherence volume. The quantum mechanical description predicts that the simultaneous response probability of the two detectors exactly vanishes, as required by the energy conservation law, because each of the two detectors needs for its activation a whole energy quantum $h\nu$. However, the classical concept predicts for this case a finite non-zero coincidence probability.

We calculate quantum mechanically and write the (undelayed) coincidence counting rate (i.e. the number of registered coincidences per second) in the following form

$$C(0) = \beta^2 T_1 \langle n(n-1) \rangle, \tag{8.11}$$

where T_1 is the detector response time and the constant β is the ratio of the detection efficiency and the length of the coherence volume rewritten as a time. (When the cross section of the mode volume is imaged only partly on the detector's sensitive surface, the detector sensitivity is decreased by an additional factor.) The factor T_1 on the right hand side of the equation reflects the fact that generally the response probability of the two detectors is proportional to the product of the lengths Δt_1 and Δt_2 of the respective time intervals during which the first and the second detector measure. To obtain the coincidence counting rate we have to divide by one of the times Δt_1, Δt_2 and the other is replaced by T_1. The counting rate of a single detector is proportional to the mean intensity (the classical and quantum mechanical descriptions coincide at this point), and hence it is given by

$$Z = \beta \langle n \rangle. \tag{8.12}$$

From this follows the *random* coincidence counting rate

$$C_{\mathrm{ran}} = Z^2 T_1 = \beta^2 T_1 \langle n \rangle^2. \tag{8.13}$$

Of particular interest is the excess in the systematic coincidence counting rate when compared with the random one. Relating it to the random counting rate, we introduce the relative excess coincidence counting rate, for which we find, using Equations (8.11) and (8.13), the expression

$$R \equiv \frac{C(0) - C_{\text{ran}}}{C_{\text{ran}}} = \frac{\langle n(n-1) \rangle - \langle n \rangle^2}{\langle n \rangle^2} = \frac{(\Delta n)^2 - \langle n \rangle}{\langle n \rangle^2}. \tag{8.14}$$

We state with delight that the detector sensitivity has disappeared from this equation. The measurement of the relative excess coincidence counting rate characteristic of the photon statistics can be realized without problems using bad detectors and, in addition, we do not have to be extremely careful with the imaging of the optical field onto the detector surface. (We only have to guarantee that the dimension of the imaged area does not exceed the transverse coherence length.) We point out that the insensitivity of the measurable parameter Equation (8.14), is lost when considering instead the parameter $[(\Delta n)^2 - \langle n \rangle]/\langle n \rangle = (\Delta n)^2/\langle n \rangle - 1$. This so-called Mandel's Q-parameter is a convenient theoretical measure for photon statistical properties, but has the drawback that it is not directly observable.

Finally, let us mention that using Equation (8.9) allows us to calculate the relative excess coincidence counting rate (Equation (8.14)) for polarized thermal light, the result being $R = 1$, in agreement with Fig. 8.4.

8.3 Random photon distribution

We can now raise the question of the photon statistical properties of radiation fields which are not of thermal type: we think first of laser light. In fact, there is a fundamental difference between the effective emission mechanism in a laser and that in a thermal radiation source. The consequence is that the photon statistics in the two cases are also completely different. Laser light exhibits, in contrast to thermal radiation, a high degree of amplitude stabilization. Let us explain this point in more detail. While the atoms of a thermal radiation source mainly radiate spontaneously, the laser action is dominated by stimulated emission due to the high electric field strength. This brings "law and order" into the emission process. The physical mechanism is as follows: the field in the laser resonator, which has built up starting from spontaneous emission, induces electric dipole moments on the atoms of the laser medium, excited by the pump process. Between them and the driving force – given by the electric field strength – well defined phase relations are established. (The phase relation is such that it enables the dipole moment to do maximum work on the field.) Because the field is coherent over the whole resonator volume, all the dipole moments oscillate with the same "beat" and a *macroscopic* polarization builds up in the medium. (This is defined as the sum of all dipole moments in the unit volume.) The laser emission is therefore a collective process.

In addition, the atoms emit their energy much faster than they do via spontaneous emission, the quantum mechanical probability for the transition from a higher level to a lower level responsible for the emission being proportional to the intensity of the field. The emitted light amplifies the inducing field; i.e. the light waves emitted by the atoms have the same frequency, propagation direction and polarization as the driving field. In this way the almost miraculous properties of laser beams – enormous spectral density, sharp frequency and propagation direction – are formed.

In fact, these properties of laser light represent "only" a qualitative progress, because even though they are overwhelming, they are obtainable – here speaks the theoretician – by making extremely hot (thermal) light sources and employing filters and diaphragms of the highest quality. Actually, the amplitude stabilization is responsible for the fundamental qualitative difference between laser light and thermal radiation.

How does this special feature of laser radiation come about? It is a mechanism specific to the laser, the so-called saturation, that is responsible. By saturation we mean the following: the excited atoms lose their energy through the process of emission, and they finally reach the lower level. They are excited again by the pump process and can radiate again etc. As mentioned above, the elementary emission process (in the form of induced emission) is faster the higher the intensity of the driving radiation. The pumping, however, has a constant rate. The result is that the mean number of excited atoms – and with it, the surplus number, decisive for the laser action, of excited atoms over those not excited, the so-called inversion of the laser medium – decreases with increasing intensity of the laser field, and this effect is called saturation.

The saturation mechanism is responsible (in stationary laser action) for the stabilization of the amplitude and the intensity, respectively. For instance, when a momentary increase of intensity compared to its stationary value occurs, it will cause a decrease in the inversion (as explained above). This will lead in turn to a weaker emission and the initial intensity peak will be attenuated. Similarly, a momentary intensity drop is damped away via the induced short-time increase of inversion.

The phase of the laser light does not have an analogous "resetting mechanism;" it changes in time in a completely random way. The linewidth of the laser radiation is (under ideal conditions) determined mainly by this phase diffusion – in distinct contrast to the situation for thermal radiation where phase *and* amplitude fluctuations contribute equally to the linewidth.

The absence of intensity fluctuations does not allow for a surplus of coincidences, and a measurement must yield a coincidence counting rate independent of the delay time. However, the existence of a constant intensity does not imply that the number $n(t; T)$ of registered photons during a time interval T also remains the same. Only the *probability* of the response of the counter is always constant. Enough freedom remains for the counting events to result in a statistical distribution of the photon numbers $n(t; T)$.

The form of the distribution is predictable once we use the correspondence between classical monochromatic waves with well defined amplitudes and phases and the quantum mechanical Glauber states discussed in Section 4.4. As already explained, in this case the photons related to the mode volume follow a Poisson distribution, which, according to Equation (4.6), takes the form (we replace the parameter $|\alpha|^2$ by the mean photon number $\bar{n}$)

$$p_n = \mathrm{e}^{-\bar{n}} \frac{\bar{n}^n}{n!} \qquad (n = 0, 1, 2, \ldots). \tag{8.15}$$

The distribution is basically different from the Bose–Einstein distribution valid for thermal light as illustrated in Fig. 8.6: it shows a pronounced maximum at a position near the mean photon number $\bar{n}$ and is much narrower than the Bose–Einstein distribution. (It is known that $\Delta n^2 = \bar{n}$ holds for the Poisson distribution.)

We identify the mode volume with the coherence volume V_{coh} in the same way as in the theoretical description of thermal radiation in Section 8.2, whereby the coherence volume V_{coh} is characterized by the property that the phase does not change significantly within it.

It is now necessary to draw conclusions about the statistics of the count events from the known statistical behavior of the photons present in the coherence volume. As explained in Section 8.2, the influence of either an observation volume smaller than the coherence volume or that of inefficient detectors can generally be described as an effective beamsplitting process. Because in such cases a Poisson distribution is transformed into a distribution of the same type, we arrive, as for thermal light, at the result that Equation (8.15) applies also to photon numbers found experimentally for integration times T shorter than the coherence time t_{coh}. The detectors need not be ideal, and the light beam cross section might be imaged only partly onto the detector surface. The Poisson distribution describes, as is known from classical statistics, completely random processes, and therefore Equation (8.15) applies also to $T > t_{\mathrm{coh}}$, i.e. it is valid for arbitrary integration times.

We can, using Equation (8.15), easily calculate the mean value of $n(n-1)$ characteristic for (undelayed) coincidences. It is a straightforward matter to state; we

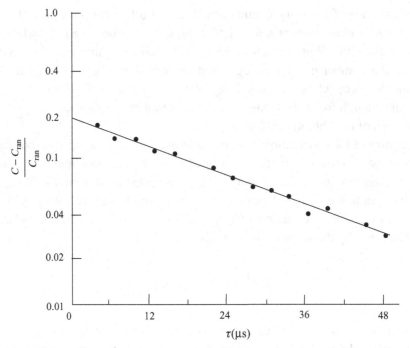

Fig. 8.7.   Relative excess coincidence counting rate with respect to the random rate C_{ran} as a function of the delay time τ, measured on a single-mode gas laser whose output power was 2.7 times the threshold power. From Pike (1970).

find the result

$$\langle n(n-1) \rangle = \langle n \rangle^2 \qquad \text{(for a Poisson distribution)}, \qquad (8.16)$$

indicating that the coincidence counting rate to be measured agrees precisely with the random one. The relative excess coincidence counting rate R introduced in Section 8.2 (Equation (8.14)) equals due to this zero. Laser radiation does not – in contrast to thermal radiation – exhibit a tendency towards photon bunching. The probability of detecting two photons (at the same position) with a time delay τ is the same for all values of τ.

All the statements about laser light made up to now refer to an idealized case which can be realized only approximately. However, the radiation of a single-mode laser operated far above the threshold comes very close to the ideal. The saturation near threshold is small, and the amplitude stabilization mechanism based on it becomes less efficient. This results in intensity fluctuations. Near threshold we observe a noticeable surplus of coincidences compared with the random case, as illustrated by Fig. 8.7.

It is interesting to follow the onset of laser action with the photon counting technique. (Fortunately the build up time of laser oscillations is quite long – for the

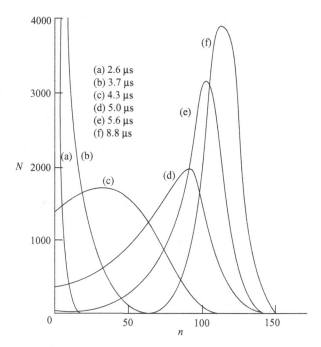

Fig. 8.8. Measured photon distribution in the build up regime of a single-mode gas laser. The plots show the number of cases N when n photons were registered for different times after switch on. From Arecchi, Degiorgio and Querzola, (1967a).

He–Ne laser, for example, it is of the order of microseconds – and an integration time short compared to this is achievable without great effort.) The result found experimentally is shown in Fig. 8.8. The photon distribution in the initial stage of the oscillation differs only slightly from a Bose–Einstein distribution; however, it becomes more and more similar to a Poisson distribution as time elapses.

8.4 Photon antibunching

From the classical point of view, by amplitude stabilization of light the utmost is done to minimize the fluctuations of the photon number (in the sense of counting events). Accordingly, the Poisson distribution represents the greatest order a radiation field might have from the classical point of view.

Quantum theory does not share this view. It declares as possible, states of the electromagnetic field which have photon distributions narrower than a Poisson distribution. Naturally, such states have no classical analog, and they are manifestations of the specifically quantum mechanical (ultimately corpuscular) aspect of the radiation.

In fact, the states with sharp photon numbers (the eigenstates of the energy operator of the free field) appearing "naturally" in the quantum formalism are of this "strange" type. Moreover, a detector with ideal efficiency responding to the field contained in the whole mode volume would not indicate any fluctuations in photon number! To be able to predict the results of a realistic measurement, we have to subject the distribution of the numbers of photons present in the mode volume to the Bernoulli transformation, Equation (7.3). In the present case of a sharp photon number n in the mode volume, Equation (7.2) describes directly the distribution of the number of *registered* photons, k. The original distribution, as expected, is considerably broadened due to the influence of statistical laws governing the "selection" of photons by the detector from the ensemble of photons contained in the mode volume. These laws act because the observed volume V_{obs} is smaller than the mode volume or the coherence volume and the detection efficiency η is smaller than unity. The corresponding variance is not difficult to calculate. According to Equation (7.4), the relations

$$\langle k \rangle = tn, \quad \langle k(k-1) \rangle = t^2 n(n-1) \tag{8.17}$$

hold, and imply

$$\langle k^2 \rangle - tn = \langle k \rangle^2 - t^2 n \tag{8.18}$$

or

$$\Delta k^2 = \langle k^2 \rangle - \langle k \rangle^2 = (1-t)\langle k \rangle. \tag{8.19}$$

The variance is greater the more the overall damping factor t (given by the product of V_{obs}/V_{coh} and η) differs from unity; i.e. the smaller the average number of registered photons. In the real experiment, there is no trace of a sharp photon number! Let us note that it is not only a shorter integration time – compared with the coherence time – and a lower detection efficiency that are responsible for such a "falsification," but an insufficient focusing of a light beam coherent over its whole cross section onto the sensitive detector surface also has the same effect. We can – apart from the cases of thermal and ideal laser radiation, as we have seen above – measure the true photon statistics only when the detector registers with certainty all the photons contained in the mode volume. Measured photon distributions, except Bose–Einstein and Poisson distributions, yield only limited information as long as the effective damping factor t is unknown.

From Equation (7.4), we can infer quite generally, i.e. for the case of arbitrary photon statistics, the following relation between the measured variance and the real variance of the photon number:

$$\frac{\Delta k^2}{\langle k \rangle^2} - \frac{1}{\langle k \rangle} = \frac{\Delta n^2}{\langle n \rangle^2} - \frac{1}{\langle n \rangle}, \tag{8.20}$$

which implies the relation

$$\frac{\Delta n^2}{\langle n \rangle^2} = \frac{\Delta k^2}{\langle k \rangle^2} + (t-1)\frac{1}{\langle k \rangle}. \tag{8.21}$$

Let us now turn our attention to coincidence measurements. As explained in Section 8.2, the relative excess coincidence counting rate R is fortunately independent of the detector sensitivity. For a sharp photon number n (in the mode volume) it takes, according to Equation (8.14), the value $R = -1/n$. Because Δn^2 can be at best zero, this is also the biggest (with respect to its modulus) negative value of R for a given *mean* photon number.

Hence for the $R < 0$ case we do not predict a surplus of the coincidences but a deficit. Because for large delay times $\tau \gg L/c$, with L being the length of the mode volume, we detect only random coincidences, the obtained result means that two photons are found within a smaller distance τ ($\leq L/c$) less often than for larger distances. We are dealing with an effect which is opposite to that characteristic of thermal light (bunching), and in the literature the term "antibunching" has been introduced for it. The photons do not seek mutual nearness, but instead prefer a certain "distance" between each other. This shows that "bunching" is not a fundamental effect. (Originally it was widely held that it is a direct consequence of the fact that photons, as spin 1 particles, have to obey Bose statistics.) In fact, it is the particular form of light generation that determines whether the photons seemingly attract or repel each other.

The appearance of a negative value of R is not comprehensible within classical optics; it is a manifestation of the particle aspect of radiation. Classically, the co-incidence counting rate is proportional to the intensity correlation, i.e. to the time averaged value of $I(t)I(t + \tau)$. Under stationary conditions this is just the auto-correlation function, and it is well known that such a function has its absolute maximum at $\tau = 0$ (see, for example, Middleton (1960)). For our situation, this implies that the coincidence counting rate must have its largest value for $\tau = 0$.

Based on the arguments given above it is clear that the "antibunching" effect is noticeable only for small mean photon numbers (related to the mode volume); it is, in fact, of microscopic nature. For $\langle n \rangle \to \infty$ it vanishes in a pleasing agreement with Bohr's correspondence principle requiring that the quantum mechanical description goes over into the classical one for high excitations (in our case for high photon numbers).

Based on Equation (8.14), we can distinguish between classical and non-classical light. The borderline is at $R = 0$, i.e. for $\Delta n^2 = \langle n \rangle$, and our results can be summarized as in Table 8.1.

It is also possible to determine the photon distribution through photon counting. Let us recall, however, that under realistic experimental conditions (in particular through the use of inefficient detectors) the distribution will be falsified: it becomes

Table 8.1. *Classification of light based on the relative excess coincidence counting rate R.*

The last column states the relation of the photon distribution with respect to the Poisson distribution.

R	Δn^2	Photon statistics	Photon distribution
> 0	$> \langle n \rangle$	"bunching"	super-Poisson
$= 0$	$= \langle n \rangle$		Poisson
< 0	$< \langle n \rangle$	"antibunching"	sub-Poisson

broadened, as can be seen from Equation (8.21). However, once we measure a distribution that is narrower than a Poisson distribution with the same mean photon number, we have proved that we are dealing with non-classical light.

The most important question is whether non-classical light really exists, or, in other words, is it possible to realize radiation fields with $\Delta n^2 < \langle n \rangle$ and so make "photon antibunching" a measurable effect?

The question may be answered "yes" in principle. The simplest idea is to use as a light source a single atom excited to resonance fluorescence by an incident laser wave (see Section 6.1). Such a system emits photons, one after the other. There is, however, a time delay between two subsequent emissions – after each emission the atom has to be excited again, which takes a finite time. Observing the radiation emitted in a certain direction, we see the photons ordered like a string of pearls. However, the distances between the photons are unequal, very short separations seldom occur, and zero separation is never observed. Coincidence measurements on such a light beam would yield none for delay time 0 (ideal conditions assumed), but there would be coincidences for longer delay times.

All this sounds easy, but a real experiment is very difficult to set up. The main obstacle is the detection of the radiation from a *single* atom. As soon as it is possible that a second atom contributes to the observed radiation, coincidences can appear because it will sometimes happen that both atoms emit one photon simultaneously, each in the same direction. The first experiment demonstrating "photon antibunching" was performed as follows (Kimble *et al.*, 1977; Dagenais and Mandel, 1978). The light source was formed by an atomic beam consisting of sodium atoms. Resonance fluorescence of the atoms was induced by a resonant laser beam (incident orthogonal to the atomic beam). The sideways emitted radiation was observed. Only a fraction of the radiation, emitted from a small section of the atomic beam of 100 μm length (the diameter of the beam was also 100 μm), was collected using a microscope objective and sent, via a balanced beamsplitter (see Fig. 8.3), to the two detectors of the coincidence counting

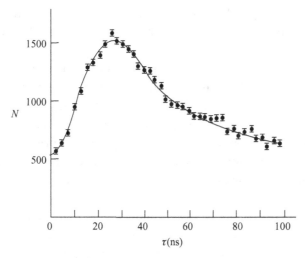

Fig. 8.9. Measured numbers of coincidences N as a function of the delay time τ in the case of resonance fluorescence. From Dagenais and Mandel (1978).

apparatus. The velocity of the atoms was chosen such that the observed volume contained, on average, less than one atom. However, it could not be ruled out that at certain times the volume contained two, or even three, atoms simultaneously. (And at other times, there was no atom present at all.) We can assume that the atoms contained in the observed volume fluctuate according to a Poisson distribution. The consequence is, as a theoretical analysis (Jakeman *et al.*, 1977) revealed, that no coincidence deficit (in the sense of the inequality $R < 0$) should be found.

In agreement with this, the results of the measurements shown in Fig. 8.9 indicate that the number of measured coincidences – as a function of the delay time τ – has a minimum at $\tau = 0$ which does not go to zero as it should for a single atom (under ideal conditions): the minimum even lies above the number of random coincidences.

The results of the experiment are not very satisfactory. However, as previously explained, the appearance of a *relative* minimum at $\tau = 0$ in Fig. 8.9 is in contradiction to classical theory and it bears witness to the particle aspect of light. In more recent experiments the required condition was satisfied and the detector "took notice" of only a single atom. This was achieved with the help of modern trapping techniques (see Section 6.1): using optical cooling it is possible to trap a single atom for as long as necessary in a radio frequency quadrupole trap (Paul trap). "Photon antibunching" was demonstrated on a trapped Mg^+ ion in complete agreement with theory (Diedrich and Walther, 1987). Another possibility is to excite a single pentacene molecule located in a p-terphenyl crystal, thus producing fluorescence (Basché *et al.*, 1992). The process was not resonance fluorescence, however;

instead, the deexcitation took place in two steps. However, the emission conditions are very similar, and the fluorescence light exhibited the "antibunching" effect in its full beauty.

One might object that the described experiments do not invalidate classical electrodynamics but just the classical assumption that the photodetector responds to the instantaneous intensity independently of how much energy is really available for absorption. In fact, the essential discrepancy between the classical and the quantum mechanical description of the discussed experiments originates from the fact that the classical formula for the simultaneous response of two photodetectors predicts observable coincidences even for single incident photons. However, this is impossible due to energetic reasons!

The classical theory cannot be saved simply by improving the description of the photoelectric detection process. The deeper reason for its failure is the fact that – as a wave theory – it cannot avoid predicting the splitting of single photons incident on a balanced beamsplitter. (Otherwise there could be no "interference of the photon with itself.") This leads, however, as discussed in detail in Section 7.1, to a contradiction with our experience. A correct description of the way in which a detector works has always to take into account that half photons cannot be absorbed. This would imply that, provided an ideal source were available, none of the detectors would respond, in the experiment by Mandel and coworkers.

The light generated by a single atom through resonance scattering has the "antibunching" property from the very beginning. We might ask whether it is possible to alter already generated light in such a way that it becomes "antibunched." In fact, a change of the photon statistics is not surprising. For example, a two-photon absorber (the individual atoms always "swallow" two photons at once) suppresses existing intensity fluctuations. This is easily seen from the classical description; the probability of absorption is in this case proportional to the squared intensity and therefore the intensity peaks are disproportionately suppressed. (The ratio of the maximum and the mean intensity decreases during interaction, in contrast to one-photon absorption where it stays constant.) The result is that the intensity becomes almost constant. The classical treatment predicts a completely amplitude stabilized field as the final stage.

A constant intensity does not imply, as was explained in the preceding text, a fixed value of the photon numbers; these will fluctuate according to a Poisson distribution. Would it not be possible to reduce further such fluctuations through interaction with a two-photon absorber?

Quantum mechanical calculations show that the conjecture is correct. It is easily seen from a naive photon picture (in which we imagine photons as energy packets localized in space). An atom absorbs two photons more or less simultaneously in any elementary act. To be more precise, we should say that the delay between the arrival of the two photons should not exceed a maximal value τ_{max} for absorption to be possible. (For simplicity we assume the incident beam to be so thin that the photons are sufficiently close to each other in the transverse direction so that one and the same atom can absorb them – provided their temporal distance is sufficiently short.) The two-photon absorber always "picks out" from the field pairs of photons that are close to each other. This leads (starting from an amplitude stabilized beam for which the probability of finding two photons with a time separation τ is independent of τ) to the formation of a deficit of photon pairs with $\tau \leq \tau_{max}$ and this is exactly what is understood by "photon antibunching." The final state for long enough interactions would be a state with photons separated in time by $\tau > \tau_{max}$ (provided we could eliminate all disturbing factors such as scattering, weak one-photon absorption through impurities, etc.).

The process faces another, insurmountable, obstacle, however – the extremely low efficiency of two-photon absorption. This implies that a considerable attenuation of the beam is possible only for high intensities – on the other hand, as can be seen from Equation (8.14), a chance to observe the "antibunching" effect exists only for very low intensities.

It would be highly desirable to have a type of two-photon process which is efficient at small intensities. In fact, there is such a process in the form of induced harmonic generation in a non-linear crystal, by which we mean a weak fundamental wave at frequency ν incident onto a crystal, together with an intense harmonic wave (oscillating at the doubled frequency 2ν). The phases of the two waves φ_ν and $\varphi_{2\nu}$ are chosen in such a way (the time independent difference $\varphi_{2\nu} - 2\varphi_\nu$ is relevant) that the fundamental wave is attenuated and the harmonic is amplified. The fundamental wave is affected by a kind of two-photon absorption because the elementary process of the interaction is a "fusion" of two photons of the fundamental wave into a photon of the harmonic according to the equation

$$h\nu + h\nu = h(2\nu). \tag{8.22}$$

It is to be expected – and quantum mechanical calculations confirm this – that the (attenuated) fundamental wave leaving the crystal exhibits the "antibunching" effect. We emphasize that the intensity of the incident wave can be arbitrarily small; the process is driven by the incident strong harmonic. Experiments of this type have not yet been achieved.

Considering electrons instead of photons, the "antibunching" effect appears as something very natural. Due to the Coulomb repulsion it is in the nature of electrons not to approach too close to each other. In fact, this type of behavior can be transferred from electrons onto photons. The task is to convert electrons – when possible in a 1:1 ratio – into detectable photons. This happens, for example, in the Franck–Hertz experiment. Here atoms are bombarded by a beam of electrons and are thus resonantly excited. Subsequently they spontaneously emit a photon each. The experimental verification of the effect was successful (Teich and Saleh, 1985): the generated photons exhibited sub-Poisson statistics. However, the measured deviation from Poissonian statistics was very small. The reason is mainly due to the fact that not each electron was definitely "converted" into a photon which then went on to be detected. The weak points of the experiment are (i) the conversion efficiency of the electrons, which is below 100%, (ii) the fact that the photons are emitted in all directions, which allows only a fraction of the photons, collected by a lens, to be directed onto the detector; and (iii) the inefficiency of the detectors.

Another way to convert electrons into photons is presented by semiconductor lasers. More precisely, it is the annihilation of electron and hole pairs which leads to light emission. One advantage compared with the Franck–Hertz experiment is that the photons are emitted in a well defined direction. On the other hand, the resonator – the end surfaces of the laser crystal act like mirrors – hinders the photons' immediate exit. This leads to a time spread of photons generated at the same instant. Because of this, a sub-Poissonian statistics of the photons can be observed only when the measurement time is longer than the mean lifetime of the photons in the resonator. In addition, it is necessary to stabilize very precisely the pump current. Finally, laser crystals of excellent quality are needed to make losses of photons due to absorption and similar processes as small as possible. In fact, the discussed effect could be observed using an InGaAsP/InP laser with distributed feedback (Machida, Yamamoto and Itaya, 1987). In practice, instead of photon counting, the noise spectrum of the photocurrent is measured. The appearance of sub-Poissonian photon statistics from a certain measurement interval T onward is indicated by a drop in the noise power spectrum below the shot noise for frequencies $v > 1/T$.

Another light source producing non-classical radiation by itself is the one-atom maser, also called the micromaser. Atoms excited into a certain Rydberg state (see Section 6.2) are sent one after the other through a microwave resonator tuned to a (microwave) transition into one of the lower lying Rydberg states. Thanks to the unusually large transition dipole moment of Rydberg atoms – a consequence of the extremely elongated electron orbits – a strong interaction between the atoms and the radiation field takes place in the presence of a few (or zero) photons in the resonator. The consequence is that stationary maser operation is possible at small mean photon numbers (Meschede, Walther and Müller, 1984). A necessary prerequisite is, however, a resonator with extremely small losses, which is achievable using superconducting materials cooled with liquid helium. The strong cooling has the additional desirable effect of suppressing the thermal radiation from the resonator walls. It is amazing that in such a maser a resonator field with sub-Poissonian statistics can be generated in the interplay between the energy supply by the atoms and the resonator losses for suitable chosen parameters (such as the time of flight of the atoms, the atomic flux, etc.). The detection of this non-classical behavior can be achieved only indirectly. There are no microwave photon counters, and we are fully dependent on concluding the photon statistics from the statistics of the de-excited atoms – with the aid of field ionization detectors it is easily possible to find out whether an atom leaves the resonator in the upper or the lower state of the maser transition. Fortunately, there is a close connection between the two statistics (proven by theory); in particular, the type of statistics (super-Poisson, Poisson and sub-Poisson) of the atoms and the photons is always the same. The corresponding experiments have been successfully performed at Garching (Rempe, Schmidt-Kaler and Walther, 1990).

Finally, let us also note that by external manipulation of previously generated light one can change its statistics and generate "antibunching." Here again we make use of parametric fluorescence (see Section 6.7). One possibility is to use the idler wave for the manipulation of the pump wave. The idler is incident on a detector and its output signal is used to close a shutter inserted into the pump beam for a short time (Walker and Jakeman, 1985). In this way, pieces of equal length are cut out from the signal wave at irregular intervals. Obviously with this procedure we decrease the probability of finding two photons within a short interval, and this is clearly what is meant by "antibunching."

9

Squeezed light

9.1 Quadrature components of light

In the previous chapter we learned about a special form of "non-classical light." It was the particle character of the light that could not be properly described within the framework of a classical field theory (classical optics). We can, however, visualize light particles – photons – when we compare them with bullets. For example, it is not difficult to imagine a sequence of fast moving particles lined up like a string of pearls – similar to the bullets fired by a machine gun in a certain direction – to visualize ideal "antibunching."

The abnormal (from the viewpoint of classical optics) photon statistics does not begin to exhaust the wealth of curiosities that Nature has at hand. It is possible to manipulate the quantum mechanical field fluctuation in a very subtle way, which leads to measurable effects, and we will concentrate on this problem in the following.

First of all, we have to introduce an alternative description of the field which is more suited to the problem under scrutiny. Usually we quantize the (classical) electric field strength by decomposing it into positive and a negative frequency parts. This results in the representation involving plane waves,

$$E(\mathbf{r}, t) = A\,e^{-i(\omega t - \mathbf{kr})} + A^*\,e^{i(\omega t - \mathbf{kr})}, \tag{9.1}$$

where A is a complex amplitude. Instead of Equation (9.1), we write the electric field strength as

$$E(\mathbf{r}, t) = X\cos(\omega t - \mathbf{kr}) + P\sin(\omega t - \mathbf{kr}), \tag{9.2}$$

with two *real* amplitudes X and P. Apart from a factor of two, these are the real and imaginary parts of the complex amplitude A. We learn from comparison with a material harmonic oscillator that the variables X and P (in proper normalization) have the meaning of position and momentum (which explains our notation). Let us

155

write Equation (9.2) in the form

$$E(\mathbf{r}, t) = C\{x \cos(\omega t - \mathbf{kr}) + p \sin(\omega t - \mathbf{kr})\}. \tag{9.3}$$

The normalization factor C can be chosen such that the (Hermitian) operators $\hat{x}$ and $\hat{p}$ corresponding to the classical variables x and p will satisfy the commutation relation

$$[\hat{x}, \hat{p}] = \mathrm{i}\hat{1}. \tag{9.4}$$

This relation differs from the well known Heisenberg commutation relation for position and momentum only in a factor of $\hbar$ missing from the right-hand side. The commutation relation implies Heisenberg's uncertainty relation, and therefore the properly normalized real electric field amplitudes will satisfy the uncertainty relation

$$\Delta x \, \Delta p \geq \frac{1}{2}. \tag{9.5}$$

The amplitudes x and p are usually called field quadrature components. Choosing a reference wave that oscillates as $\cos(\omega t - \mathbf{kr})$, we can interpret x and p as the in phase and out of phase quadrature components.

What is the meaning of the inequality in Equation (9.5)? First of all, it states that both quadratures necessarily fluctuate. It is natural to ask what those quantum states that have the minimal uncertainty that is compatible with quantum mechanics look like, i.e. for which the equality sign holds in Equation (9.5). This problem was solved in 1933 by W. Pauli (1933) in his famous *Handbuch der Physik* in an elegant three-line proof. The result is that the *unique* solutions are Gaussian wave functions,

$$\Psi(x) = (2\pi)^{-\frac{1}{4}}(\Delta x)^{-\frac{1}{2}} \exp\left\{-\frac{(x - \langle x \rangle)^2}{(2\Delta x)^2} + \mathrm{i}\frac{\langle p \rangle x}{\hbar}\right\}, \tag{9.6}$$

where Δx is the square root of the variance of x, i.e.

$$\Delta x = \sqrt{\langle (x - \langle x \rangle)^2 \rangle}.$$

Pauli was thinking of x and p as the position and the momentum of a particle; however, we can equally well interpret these variables as quadrature components, and there is nothing to stop us assigning a Schrödinger wave function $\Psi(x)$ to the quantum state of the field. We simply choose the representation with respect to the quadrature component x instead of the usual expansion in terms of Fock states!

It is important to point out that the (positive) parameter Δx was chosen *arbitrarily* in Equation (9.6). This equation thus describes a whole class of quantum mechanical states of the radiation field. Among them, those with equally strong fluctuations in both quadrature components are distinguished. For the quadratures, then, the relation $(\Delta x)^2 = (\Delta p)^2 = 1/2$ holds. It seems that this is a natural symmetry property of light, and the coherent states described in Section 4.4 have this property. Also, the fluctuations of x and p for these states are exactly the same as for the vacuum state (which can be considered as the limiting case of a coherent state with an infinitely small complex amplitude α).

In general, the states in Equation (9.6) have the property that the fluctuations differ for the two quadrature components. Due to the validity of the relation $(\Delta x)^2 (\Delta p)^2 = 1/4$, the variance of one of the quadratures is larger than the vacuum value $1/2$ and the other is smaller. At first glance, this statement seems astonishing since we consider the vacuum fluctuations to form a natural lower limit. However, when we analyze the problem in the context of a harmonic oscillator, we do not find it surprising that one can localize a particle with a greater precision than that corresponding to its ground state. On the contrary, it is one of the axioms of quantum mechanics that we can measure the particle's position with arbitrary precision. Why then should it be impossible for light to have quadrature components, one of which is better defined than for the vacuum state?

The most important question concerns the practical realizability of such light, which soon received its own name – "squeezed light." This was a challenge for theoreticians as well as experimentalists. Apart from the question of how to generate such a strange form of light, it was also a problem to detect its squeezing properties. The fact that this task could be successfully accomplished is one of the greatest achievements in quantum optics. We will discuss the steps taken to achieve this in the following sections.

9.2 Generation

Processes associated with non-linear optics are best suited for the generation of squeezed light. Especially efficient is degenerate parametric down-conversion: a special type of three-wave interaction (as discussed in Section 6.7), where the signal and the idler wave coincide. Let us consider the case of a strong pump wave and a weak signal wave. The underlying physical mechanism is the formation of a polarization of the medium by the pump wave (we neglect the depletion of the pump due to amplification of the signal) and the signal wave, and this polarization drives in turn the signal wave. The complex amplitude A of the signal wave satisfies the equation of motion

$$\dot{A} = \kappa A^*, \tag{9.7}$$

where the effective coupling constant κ is proportional to the non-linear susceptibility of the medium and the complex amplitude of the pump wave. In the following we will choose the phase of the signal wave such that κ will be positive.

Equation (9.7) can be used to write down the equations of motion for the quadrature components x and p, which are (apart from a normalization factor) the real and the imaginary parts of A:

$$\dot{x} = \kappa x, \tag{9.8}$$

$$\dot{p} = -\kappa p. \tag{9.9}$$

Obviously they describe an exponential *increase* in x and an exponential *decrease* in p. The process of amplification will cause one quadrature component to be enhanced, whilst at the some time the other will be attenuated. This asymmetry is also transferred to the uncertainties of Δx and Δp, for which Equations (9.8) and (9.9) result in

$$(\Delta x)_t^2 = e^{2\kappa t}(\Delta x)_0^2, \tag{9.10}$$

$$(\Delta p)_t^2 = e^{-2\kappa t}(\Delta p)_0^2. \tag{9.11}$$

This behavior is exactly what we have been looking for, namely a dramatic decrease in one of the uncertainties. The increase of the other uncertainty is the price we have to pay. (That in our treatment the p quadrature component is "squeezed" is the result of the special choice of the signal phase. If we changed it by π, the x component would be squeezed.)

Since the equations of motion are linear, the performed analysis is valid in the classical as well as in the quantum case, and hence, with the degenerate parametric amplifier, we have an instrument at hand which enables us to prepare squeezed states from coherent (Glauber) states. The variances of the initial Glauber state $(\Delta x)_0^2$ and $(\Delta p)_0^2$ equal the vacuum value $1/2$ and, according to Equation (9.11), one of the quadratures will be more and more suppressed during the amplification process.

The described squeezing experiment was performed successfully (Wu *et al.*, 1986). The experiment did not start from a coherent but from the vacuum state. The process, which is also called subharmonic generation, is initiated by the parametric fluorescence discussed in Section 6.7. The weak signal thus produced is then further amplified. The resulting state is usually termed "squeezed vacuum."

The credit for generating and detecting squeezed light for the first time, however, goes to another American research group (Slusher *et al.*, 1985; Slusher and Yurke, 1986). They used a four wave mixing type interaction and sodium vapor as the non-linear medium. In the experiment, the reflection of a strong laser wave was utilized to create a second, counter-propagating wave. These two waves generated in a resonator – also from the vacuum – two signal waves, differing slightly in their frequencies. It was proved that the total field formed from these two waves – after their exit from the resonator – has the predicted squeezing properties. Before examining the detection technique used more closely, let us comment on the theoretical analysis achieved so far.

It follows from Equations (9.10) and (9.11) that the uncertainty product $\Delta x \, \Delta p$ is time independent. Consequently, when Heisenberg's uncertainty relation is originally fulfilled with the equality sign, it remains so for ever. According to Pauli's proof mentioned in Section 9.1, the Schrödinger wave function of the field state is always of the form in Equation (9.6), and only the parameter Δx is time dependent according to Equation (9.10). This relation also makes it clear that strong squeezing – in our case, of the quadrature component p – is connected with a significant increase in energy: when Δp becomes very small, Δx must quite generally grow at least inversely since otherwise Heisenberg's uncertainty relation, which is fundamental for quantum mechanics, would be violated. The energy of the field is proportional to the sum $x^2 + p^2$, and this means that the energy fluctuations (and therefore the mean energy) will become bigger. In contrast to photon statistics, where the non-classical behavior is easy to observe for small intensities, the non-classical squeezing effect requires high intensities to become well pronounced. A more detailed theoretical analysis shows in addition that the fluctuations are much larger than for thermal light. In other words, we are facing a "super-bunching" of photons. This is particularly true for the abovementioned squeezed vacuum. The name given to this state is not a very appropriate one as it suggests the existence of a new type of vacuum. We are not dealing here with a vacuum in the sense of a field state without photons!

The squeezed vacuum is extraordinary also with respect to its photon distribution. Because the elementary process of subharmonic generation is the conversion of a pump photon into two signal photons, only an even number of photons can be generated (under ideal experimental conditions). The photon number distribution has a comb-like structure: it is zero for all odd photon numbers. It is clear that such a "pathological" state must be very fragile for even the smallest disturbances. The comb-like structure will be completely destroyed, for example, by a single absorbing atom which interacts with the field during a time interval that is just long enough that it becomes excited with a probability of 50%.

This "fragility" is a general feature of "squeezed states" and makes their use for optical data communication rather illusory as we cannot avoid damping occurring in the optical fiber. The advantage of such an application would be to imprint the signal onto the squeezed component, thus improving considerably the signal to noise ratio. Squeezed states might possibly assume practical importance in high sensitivity interferometers. Their sensitivity with respect to mirror displacements could be increased by preventing the intrusion of vacuum fluctuations into the unused input port which "contaminate" the interferometer. To this end, squeezed light (with a proper phase) might be coupled into the interferometer instead of the vacuum. The increased effort would probably pay off in gravitational wave interferometry where a mirror displacement indicates the presence of a gravitational wave. Since inconceivably small displacements – by orders of magnitude smaller than the proton radius – are to be measured, any procedure leading to an improvement of the signal to noise ratio without additional intensity increase is more than welcome.

9.3 Homodyne detection

Radio engineers developed a very efficient detection method that comprises the mixing of a (weak) signal with a wave generated by a local sender. This technique – depending on whether the signal frequency coincides with that of the "local oscillator" or not being called homodyne or heterodyne measurement – also found an application in optical spectroscopy. A special type of optical homodyne detection, balanced homodyne detection, proved itself to be an almost ideal quantum optical measurement procedure. It is especially well suited for the detection of squeezing effects, and therefore we will examine it more closely.

The first step is the optical mixing of the usually weak radiation to be analyzed (in the following called the signal) with an intense laser wave as the local oscillator (Fig. 9.1). For the mixing we will use a beamsplitter with a high enough accuracy to achieve splitting in a 1:1 ratio. The outgoing waves impinge onto two separate detectors. The measured signal is given by the difference between the two photocurrents.

The theoretical analysis of this measurement method shows that the measured variable – apart from a factor proportional to the laser field amplitude – is just one of the quadrature components of the field (see Section 9.1). It is assumed that an intense laser wave is used, allowing for a classical description. It is especially important that we use a 50%:50% beamsplitter, as it must guarantee exact compensation of the coherent contributions of the laser field in the difference between the photocurrents, which would otherwise be dominant.

Which quadrature component will be measured? It is obvious that it will be determined by the relative phase Θ between the signal and the local oscillator. The theoretical treatment shows that we measure the x quadrature component when Θ equals zero, whereas for $\Theta = \frac{\pi}{2}$ we measure the p component. We point out that experimentally the desired relative phase can be adjusted without great difficulty. We start from a laser field and split it into two parts which are then used twofold: first (if necessary after a frequency transformation, usually frequency doubling with the help of a non-linear crystal) for the signal generation (in a non-linear

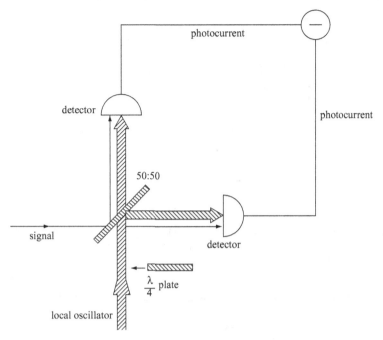

Fig. 9.1. Homodyne measurement of quadrature components x and p, respectively. The difference of the two photocurrents delivered by the two detectors is proportional to x, and, with an inserted quarter wave plate, it is proportional to p.

optical process), and second as a local oscillator. There is a certain phase relation between the signal wave and the local oscillator, and the relative phase can be easily adjusted to the prescribed value by varying the path difference. The phase fluctuations of the primary laser field obviously do not play a role. We point out that this procedure is the only feasible way of performing the homodyne measurement. The other possibilities are (i) a signal with the property that the measured data (in the sense of distributions) do not depend on Θ, or (ii) a laser with extremely high frequency stability to guarantee the (absolute) phase stability of the local oscillator during the whole time of the measurement (on an ensemble).

Employing the balanced homodyne detection technique, the measurement of x and p is simple. Once we have arranged the experimental setup for the measurement of the x component, it is sufficient to insert a $\lambda/4$ wave plate into the local oscillator path to measure the p component. This is indeed an experimental "comfort", almost without parallel; just think of quantum mechanics where measurements of the position and momentum each require a different experimental setup.

The capabilities of the balanced homodyne technique are not exhausted by this; we can, for example adjust the value of Θ. What is measured then is a kind of

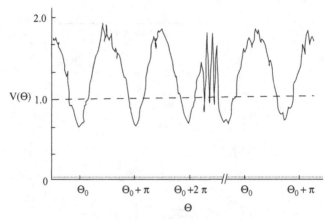

Fig. 9.2. Squeezed vacuum state detection. Relative signal from a spectrum analyzer is shown plotted against the phase Θ of the local oscillator. From Wu *et al.* (1986).

mixture between x and p, namely a new observable

$$x_\Theta = \cos \Theta\, x + \sin \Theta\, p \tag{9.12}$$

(see Section 15.6). If we measure the distributions $\omega_\Theta(x_\Theta)$ for a complete set of variables x_Θ, where Θ changes stepwise – in the sense of an approach towards a continuous progression – between 0 and π, we obtain, as discussed in detail in Section 10.3, essentially the complete quantum mechanical information about the respective system.

Let us return to the experimental detection of the squeezing effect. Based on the given arguments, we need to do the following: construct the measurement setup as shown in Fig. 9.1 and block the signal at the beginning. We cannot prevent the vacuum from entering the setup, and under these circumstances we will measure the vacuum fluctuations in the form of fluctuations of a chosen quadrature component of the field. There is no preferred phase of the electric field strength in the vacuum, and therefore the uncertainties indicated by a spectrum analyzer,[1] i.e. the square roots of the square fluctuations, are independent of the relative phase Θ of the local oscillator. They mark the "zero level." When we send the squeezed light into the measurement apparatus, the measured signal varies as a function of Θ with period π between maxima rising above the "zero level" and (compared with the maxima) less pronounced minima falling below it. The minimum is

---

[1] Under ideal experimental conditions, the signal of a spectrum analyzer is proportional to the square root of a squared quadrature component x_Θ averaged over a certain time interval. When the mean value of x_Θ itself vanishes, as is the case for the vacuum as well as for the "squeezed vacuum," the uncertainty Δx_Θ will be directly indicated. The measured signal will actually deteriorate due to the detector inefficiency and, for amplification of the individual photocurrents, the amplification noise.

exactly what we have been looking for. Such a minimum indicates that the fluctuations of a certain quadrature component x_Θ – the value of Θ does not play a role as it is related to the path difference between the signal and the local oscillator, which is not measured – are smaller than those of the vacuum; i.e. we have "squeezed" light. This result was obtained in the pioneering work on squeezing by Slusher *et al.*, (1985). It was later confirmed by other researchers using methods of light generation other than the abovementioned degenerate parametric amplification (Wu *et al.*, 1986). A typical measurement curve is shown in Fig. 9.2.

Let us emphasize the non-classical character of squeezed light. Within a classical theory it is not possible to understand the falling below the "zero-level." From the classical point of view, this level is simply the shot noise of the detector, which is a consequence of the "grainy" character of the electrons. That the photocurrent fluctuates even for constant incident light can be understood classically in the following way. The sensitive surface of the detector is formed from myriads of atoms, out of which, in a short time interval, only a few are ionized and hence the process is governed by chance. Since the ionization probability – for a uniformly illuminated detector surface – is the same for all atoms, and since each individual process is statistically independent of all the others, we find all the conditions for a Poissonian statistics to be satisfied. This implies that electrons released in an *arbitrarily* chosen time interval follow this type of statistics. The result is white (frequency independent) noise of the photocurrent, called shot noise. From this follows the fact that the shot noise represents an absolute lower limit: when we illuminate with light that is not amplitude stabilized, i.e. it exhibits intensity fluctuations, the noise can only increase and never decrease.

It was later found that the balanced homodyne measurement is also well suited to quantum mechanical measurements of the phase of light. Problems related to phase have been of interest since the early days of quantum mechanics, and hence will be discussed in some detail in the following chapter.

10

Measuring distribution functions

10.1 The quantum phase of light

The proper description of the phase of a quasimonochromatic light field, idealized as a single-mode state of the radiation field, is among the most delicate problems in quantum theory.[1] Generations of theoreticians have tried in vain to find a phase operator which would satisfy the quantum mechanical standard. The problem is the required hermiticity. Only by means of a mathematical trick was it possible to construct a strictly Hermitian phase operator. The clue was to curtail artificially the "playground" of the phase operator; i.e. to limit the dimension of the corresponding Hilbert space to a very large but *finite* value. If we study the history of the subject, we see that work by F. London (which is largely forgotten now) provided a satisfactory approach to the phase problem. London (1926) introduced the so-called phase states, i.e. quantum mechanical states with sharp phase value φ, of the form

$$|\varphi\rangle = \frac{1}{\sqrt{2\pi}} \sum_{n=0}^{\infty} e^{in\varphi} |n\rangle, \tag{10.1}$$

where $|n\rangle$ are the states with sharp photon number n (see Sections 4.2 and 15.1).

The states in Equation (10.1) account for the same internal relation between time and phase as is known from classical theory. Classically we can find the phase of a monochromatic oscillation by observing the zero passages. The oscillation variable (displacement or amplitude of the electric field strength) depends only on the combination $\omega t - \varphi$, which implies that the time evolution is equivalent to a phase shift. Such a property also has the phase state in Equation (10.1) because the time evolution of the states $|n\rangle$ with sharp photon number – and hence with

---

[1] To avoid misunderstandings, we emphasize that we are talking about the absolute phase; i.e. the phase with respect to an ideal reference wave and not a relative phase determined by a path difference and measurable by interferometric arrangements.

sharp energy $n\hbar\omega$ – reduces to a multiplication by the phase factor $\exp(-in\omega t)$. The fact that the phase states are not normalized is not worrying. On the contrary: because we expect the phase to be a *continuous* variable, the corresponding eigenstates cannot be normalizable – just think for comparison of a running plane wave which represents a momentum eigenstate. However, we should require normalizability in the sense of Dirac's delta function, since quite generally eigenvectors corresponding to different eigenvalues of a Hermitian operator should be mutually orthogonal. This is indeed a problem because the states in Equation (10.1) do not satisfy this condition, with the consequence that they cannot be the eigenfunctions of a Hermitian operator, i.e. of a well behaved phase operator. This problem is insurmountable. The phase seen as a quantum mechanical operator plays an extra role, and we might ask whether there is a special reason for this. The likely answer is that the coupling between the radiation field and the atomic system takes place via the electric field strength in which the phase *and* the real *amplitude* become fused. As a consequence, it is difficult to conclude anything about *only* the phase (or amplitude) from measured data.

The introduction of phase states, as in Equation (10.1), is essentially satisfactory. They yield a decomposition of the unit operator (and constitute a complete set of states). This allows us to interpret the squared absolute value of the scalar product of an arbitrary field state $|\psi\rangle$ with a phase state $|\varphi\rangle$ – the projection of the field state onto the phase state – multiplied by $d\varphi$ as the probability of finding in a phase measurement a value within the interval $\varphi \ldots \varphi + d\varphi$. This gives us a clear "recipe" for the calculation of phase distributions. Unfortunately, no one can tell us what such an ideal phase measurement should look like! Facing such an unsatisfactory theoretical situation, the pragmatist can only try to turn the tables: instead of starting from a profound theoretical analysis, we need to produce a practicable phase measurement strategy and only afterwards try to find a proper theoretical framework for it. This path towards an operational phase definition was taken by L. Mandel and coworkers, and we will follow their arguments in the following sections (Noh, Fougeres and Mandel, 1991, 1992).

10.2 Realistic phase measurement

We start with a classical description and write the electric field strength of a running plane wave in the form

$$E(z, t) = E_0 \cos(\omega t - \varphi - kz). \tag{10.2}$$

Using the addition theorem for the cosine and comparing the result with Equations (9.2) and (9.3), we find a simple relation between the phase φ and the quadrature components x and p:

$$\cos \varphi = \frac{x}{\sqrt{x^2 + p^2}}, \qquad \sin \varphi = \frac{p}{\sqrt{x^2 + p^2}}. \qquad (10.3)$$

It is interesting that already within classical optics the measurements of two variables are required to find the phase. The experimental procedure will exploit a balanced beamsplitter which splits the optical ray into two parts: on one will be measured the x component and on the other the p component (the balanced homodyne detection discussed in Section 9.3 is well suited for this task).

We may ask now whether this procedure is also practicable for a quantum mechanical phase measurement. A theoretician will have serious objections. The variables x and p, according to the principles of quantum mechanics, cannot be precisely measured simultaneously because they are canonically conjugated and therefore have to satisfy Heisenberg's uncertainty relation, Equation (9.5). Hence, the theoretician declares that we are not measuring the true quadrature components x and p in the described experiment but modified variables, which contain, in addition to x and p, noise components originating from the need to use a beamsplitter to be able to measure the two variables on the *same system*. The noise enters the system in the form of vacuum fluctuations via the unused port (see Section 7.6, Fig. 7.8). We can describe the situation in the following way: we can measure x and p simultaneously, but at the expense of precision. This applies in general to any pair of conjugated variables. What we really measure in the above experimental setup is not the "ideal" but a "noisy" phase. However, we can measure in this way at least something which comes close to the phase, and, in the case of large amplitudes, when the role of vacuum fluctuations becomes negligible, even coincides with it. We can formulate also an "operational approach" to the phase, in the discussed experimental setup this results in the definition of a physical variable termed *a* phase rather than *the* phase.

Let us now analyze the described (indirect) phase measurement from the quantum mechanical point of view. Because a single measurement does not tell us very much – the phase might have, for example, the particular value 0.72314 – we are interested mainly in statistical statements about the phase. We will have to perform many single measurements on an ensemble of identically "prepared" systems; for example, a sequence of ultrashort light pulses. It then seems natural to calculate from the measured data averaged values of the type $\overline{\sin \varphi}$, $\overline{\cos \varphi}$, $\overline{\cos^2 \varphi}$, $\overline{\sin^2 \varphi}$, $\overline{\sin \varphi \cos \varphi}$, etc. To extract the full information contained in the measured data we would also need to calculate, in principle, all higher moments of the form $\overline{\sin^k \varphi \cos^l \varphi}$, which is difficult to achieve practically. It is more advantageous to calculate an individual phase value φ from the directly measured data x and p using Equations (10.3), thus (on repeating the procedure on many ensemble members) determining a distribution, a so-called histogram, of the phase values. (Such a phase distribution allows us to calculate any moment $\sin^k \varphi \cos^l \varphi$.) By proceeding in this way, information about the (real) amplitude and its correlation with the phase is lost. The most advantageous method of evaluating the measured data is to interpret the measured pairs x and p as points in a phase space and then to determine their distribution – a "mountain" rising above the x, p plane. The phase space distribution $w(x, p)$ contains all the information about the light field obtainable from the described measurement device. When we are interested only in the phase, we just have to average over the amplitude. To this end, we rewrite the distribution function in polar coordinates ρ, φ, and the phase distribution $w(\varphi)$ is obtained using Equations (10.3) as the integral

$$w(\varphi) = \int_0^\infty \rho \, d\rho \, w(x = \rho \cos \varphi, p = \rho \sin \varphi). \tag{10.4}$$

When we average over the phase, we find the amplitude distribution as the marginal distribution of $w(x, p)$. Also, knowing $w(x, p)$ allows us to calculate mean values of arbitrary variables dependent on amplitude *and* phase, in particular of correlations between them. The adequate theoretical description of the measurement apparatus is to establish a general relationship between the phase space distribution $w(x, p)$ and the quantum mechanical state of the analyzed light field. Surprisingly, a very simple relation was found in the case of a strong local oscillator (this was not so in the Mandel experiments (Noh *et al.*, 1991, 1992), which made the theoretical analysis rather complicated): the directly measurable distribution function $w(x, p)$ is obtained by projecting the field state $|\psi\rangle$ onto a Glauber state $|\alpha\rangle$ (Freyberger, Vogel and Schleich, 1993; Leonhardt and Paul, 1993a); i.e, the following relation holds:

$$w(x, p) = \frac{1}{\pi} |\langle \psi | \alpha \rangle|^2, \tag{10.5}$$

where we have to identify $\alpha = x + ip$. The right-hand side is known as the Husimi or Q function, and has long been a very popular function among theoreticians. Usually it is introduced in such a way that in the expression $|\langle \psi | \alpha \rangle|^2$ the complex amplitude α is replaced by $(x + ip)/\sqrt{2}$. Equation (10.5) then takes the form

$$w(x, p) = 2Q(\sqrt{2}x, \sqrt{2}p). \tag{10.6}$$

Represented graphically, the Q-function gives us a "global" impression of the respective state. In particular, the squeezing properties of light can be immediately recognized. However, the Q-function was mostly considered as a theoretical construction. The above proof that it can be directly measured has considerably increased its physical importance. Indeed, there have been indications that other realistic phase measurement methods will lead to the Q function too (for details see below), but this has remained up to now only theory.

It is easy to see that the transition from the quantum mechanical state $|\psi\rangle$ to the Q-function is connected with information loss. This is in full agreement with the statement above that a simultaneous measurement of x and p must necessarily be inaccurate. The theoretical description indeed shows that the finer details of the quantum mechanical state $|\psi\rangle$ are no longer in the measured data. To realize this it is favorable to characterize the field state by the Wigner function rather than by the wave function (or the density matrix).

First, let us say a few words about the Wigner function or the Wigner distribution. The motivation for its introduction was the desire to find a quantum mechanical description similar to that in classical statistical physics. In statistical physics our knowledge about the system is represented by a distribution function for position *and* momentum (imagine, for simplicity, a particle with a single degree of freedom). The translation of this concept into quantum mechanics already seems hopeless due to the fact that the position (x) and the momentum (p) are not simultaneously measurable. It is well known that the wave function depends exclusively on either x or p and contains nevertheless *all* the information about the system. E. Wigner showed, however, that it is possible to define a *formal* quantum mechanical analog to the classical distribution function. This distribution, $W(x, p)$, bears his name and has the following properties.

(a) It is real, and is usually not just positive; it can also become negative.
(b) The following two marginals:

$$w(x) = \int\limits_{-\infty}^{\infty} W(x, p)\,\mathrm{d}p, \tag{10.7}$$

$$w(p) = \int\limits_{-\infty}^{\infty} W(x, p)\,\mathrm{d}x, \tag{10.8}$$

are the *exact* quantum mechanical distributions of position and momentum.
(c) The quantum mechanical expectation value of a function $F(\hat{x}, \hat{p})$ depending on position and momentum operators can be, when *symmetrically ordered*, calculated as the classical mean value of the variable $F(x, p)$, whereby the Wigner function plays the role of the classical weight function; i.e. the following relation holds:

$$\langle F(\hat{x}, \hat{p}) \rangle = \int\limits_{-\infty}^{\infty} \int\limits_{-\infty}^{\infty} F(x, p)\, W(x, p)\,\mathrm{d}x\mathrm{d}p. \tag{10.9}$$

(d) The Wigner function is normalized:

$$\int\limits_{-\infty}^{\infty} \int\limits_{-\infty}^{\infty} W(x, p)\, dx dp = 1. \tag{10.10}$$

(e) It is calculated from the wave function $\psi(x)$ or the density operator $\langle x|\rho|x'\rangle$ of the state according to the following prescription:

$$W(x, p) = \frac{1}{\pi} \int\limits_{-\infty}^{\infty} \exp(2ipy)\psi(x - y)\psi^*(x + y)\, dy \tag{10.11}$$

or

$$W(x, p) = \frac{1}{\pi} \int\limits_{-\infty}^{\infty} \exp(2ipy)\langle x - y|\rho|x + y\rangle\, dy. \tag{10.12}$$

Equations (10.11) and (10.12), which are simply Fourier transformations, are easily inverted, and so it is evident that the Wigner function contains the full quantum mechanical information about the respective state. This makes it clear that a description similar to the classical one is possible; however, there are two important limitations: due to property (a) the Wigner function cannot be considered as a true probability distribution (it is more appropriate to use the term quasiprobability distribution); property (c) makes life hard for us because we have to transform the operator function $F(\hat{x}, \hat{p})$ into a symmetrically ordered form before we can apply Equation (10.9). Because the square of a symmetrically ordered function is not, in general, a symmetrically ordered function, the calculation of higher moments – exceptions are (due to property (b)) only moments of x or p – is much more complicated than in classical statistics. The additional terms appearing due to the symmetrization express quantum mechanical corrections.

Despite these "defects," which remind us that the quantum mechanical description cannot be reduced to a classical one, the Wigner function is a very useful construction for theoretical analysis. Because it is real, it can easily be graphically displayed, and we can get a visual impression of the respective quantum state whereby relevant physical properties, like, for example, squeezing, are immediately recognized as characteristic asymmetries.

For our purposes, the relation between the Q function and the Wigner function is of particular interest. It turns out that the Q function can be obtained from the Wigner function by a convolution with a Gaussian function:

$$Q(x, p) = \frac{1}{\pi} \int\limits_{-\infty}^{\infty} \int\limits_{-\infty}^{\infty} W(x', p') \exp\left\{-\left[(x - x')^2 + (p - p')^2\right]\right\}\, dx' dp'. \tag{10.13}$$

This operation is simply a smoothing of the Wigner function, and therefore it is accompanied by a loss in the finer details of its structure (for an example, see Fig. 10.1). We have seen that the Mandel experiments lead to the Q function, and we now have a simple picture of how the unwanted noise entering the apparatus through the unused port of the beamsplitter distorts the measured data. Quantitatively, the influence of the noise can be clearly described: when we calculate the uncertainty of x and p using the Q function as a distribution function, we obtain the uncertainty relation

$$\Delta x \Delta p \geq 1. \tag{10.14}$$

The right-hand side of the relation is twice the value for Heisenberg's uncertainty relation, Equation (9.5). The analyzed realistic measurement apparatus contributes the same amount of noise to the uncertainty as the "quantum noise." The phase distribution obtained from the measured distribution function $w(x, p)$ or the Q function according to Equation (10.4) is, as a rule, always broader than the "ideal" phase distribution $w_{\mathrm{id}}(\varphi)$, which can be calculated using the phase states $|\varphi\rangle$ (see Equation (10.1)) for a given light field state $|\psi\rangle$ according to the simple formula $w_{\mathrm{id}}(\varphi) = |\langle\varphi|\psi\rangle|^2$, but for which up to now a measurement prescription is not known.

Finally, let us point out that the smoothing concept is also applicable to another noise source. This source is the additional loss of measurement precision due to photodetector inefficiencies. A non-ideal detector can be modeled as an ideal detector with an additional absorber, say a partially transparent mirror, placed in front of it. The latter again allows noise to enter, and it comes as no surprise that this additional disturbance can be described by further smoothing of the Q function (with a Gaussian function whose width is determined by the detection efficiency). Because performing two convolutions with Gaussian functions one after the other is equivalent to a convolution with a single Gaussian function, we come to the following conclusion. Taking into account the deviation of the detection efficiency η from unity, the measurement is not described by the Q function but by the stronger smoothed function – the so-called s-parametrized quasiprobability distribution

$$W(x, p; s) = \frac{-1}{\pi s} \int\limits_{-\infty}^{\infty} \int\limits_{\infty}^{\infty} W(x', p') \exp\left\{\frac{1}{s}\left[(x - x')^2 + (p - p')^2\right]\right\} \mathrm{d}x'\mathrm{d}p',$$

$$\tag{10.15}$$

where the parameter s is related to the detection efficiency as follows: $s = -(2 - \eta)/\eta$ (Leonhardt and Paul, 1993b). (Note that the Wigner function corresponds to $s = 0$ and the Q function corresponds to $s = -1$.) The measured

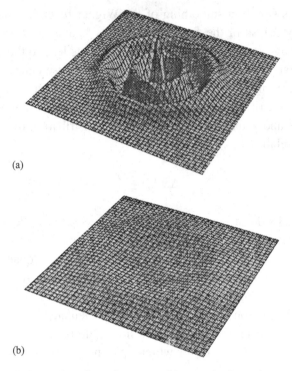

(a)

(b)

Fig. 10.1. (a) Wigner function of a state with exactly four photons and (b) the corresponding Q function obtained by smoothing it.

distribution function $w(x, p)$ is given by

$$w(x, p) = \frac{2}{\eta} W(\sqrt{2}\eta^{-\frac{1}{2}}x, \sqrt{2}\eta^{-\frac{1}{2}}p; -(2 - \eta)/\eta). \qquad (10.16)$$

The stronger smoothing in Equation (10.15) leads to an additional loss of detailed information about the field state. In particular, the experimentally determined phase distribution is broadened.

After the analysis of the Mandel scheme, let us comment briefly on earlier quantum mechanical proposals for phase measurement. The oldest relied on a simple physical idea (Bandilla and Paul, 1969; Paul, 1974). Because it is impossible to measure directly the phase on a microscopic field, we should amplify the field to macroscopic intensities, introducing as little noise as possible, with the help of a laser amplifier. The phase measurement (and an amplitude measurement if desired) on the amplified signal can be "easily" performed using classical methods. It is obvious that these methods also suffer from reduced measurement precision due to the basically unavoidable amplifier noise.

Another proposal from Shapiro and Wagner (1984) uses a heterodyne measurement. The signal is optically mixed with a strong coherent reference wave frequency shifted by $\Delta \nu$ and is detected by a photodetector. The photocurrent contains an alternating current component oscillating at the frequency $\Delta \nu$. The amplitudes of the components oscillating as $\cos(2\pi \Delta t)$ and $\sin(2\pi \Delta t)$ are measured on this current, using a quadrature demodulator. These amplitudes prove to be proportional to the quadratures we denoted above by x and p. These quantities are also measured as noisy variables, and the evaluation of the data is the same as for the Mandel scheme. But through which door does the noise now enter? The answer is: the beamsplitter mixes the signal wave not only with the strong reference wave but also, in principle, with all other vacuum oscillations. When the frequency of the reference wave is ν_0 and that of the signal is $\nu_0 + \Delta \nu$, we also have to take into consideration the wave oscillating at the "mirror" frequency $\nu_0 - \Delta \nu$ because this will also contribute (thanks to the mixing with the reference wave) to the photon current component oscillating at $\Delta \nu$ and therefore will play the role of the unwanted noise source.

When we analyze theoretically both realistic phase measurement strategies, we arrive at the rather surprising result that they lead again to the Q function. (In the case of amplification, the measurement results in an "inflated" Q function that originates from the Q function for the initial signal simply by scaling.)

In summary, we can say that all three methods are physically *completely equivalent*, and therefore the experimentally obtained phase distributions are *identical*. This is by no means obvious. A qualitative agreement in the sense that the measurement result is noisy is clear. However, the fact that different noise sources result in quantitatively identical effects is surprising. A closer inspection reveals that the noise sources, when treated formally, show a close similarity: the corresponding fluctuation operators (Langevin forces) appearing in the quantum mechanical equations of motion have the property that they guarantee the validity of the commutation relations for the creation and annihilation operators of the field (see section 15.1) after the interaction, and thus make the quantum mechanical description consistent.

We can state that the known schemes of phase measurement can be cast into a single universal scheme: generally a simultaneous measurement of two canonically conjugated variables (such as position and momentum) is performed. In this way, we measure directly (using ideal detectors) the Q function of the radiation field, and the phase distribution is obtained by averaging over the field amplitude. The price to pay is a decreased measurement precision compared with the "ideal measurement." The precision decreases further when measurements are performed using non-ideal detectors. The influence of detrimental low detection efficiency is negligible in the amplification scheme – we measure on a macroscopic object – and from a practical point of view we should give preference to this method.

10.3 State reconstruction from measured data

We demonstrated in Section 10.2 that apparatus designed for phase measurement can be used for measuring directly certain quasiprobability distributions – the Q function or a smoothed Q function when inefficient detectors are employed. The question arises as to whether with this we already have the *complete* information about the quantum mechanical state of the field to hand. The theoretician usually shows no hesitation in replying "yes." Indeed, as can be shown rather generally, this answer is correct; however, a very important assumption must be made. The respective quasiprobability distribution, for example the Q function, must be known with *absolute precision*; i.e. it must be given to us as a mathematical function. In principle, we can reverse any convolution when the smoothing function is precisely known. The smoothing of the Wigner function, Equation (10.13), does not eliminate the finer details of it *completely*; they are only strongly suppressed. The practical problem is that the measurement must be performed with extreme precision, and hence enormous experimental efforts must be made to reconstruct the Wigner function (from the experimentally determined probability distribution), thus gaining access to the full information about the quantum mechanical state. We have to conclude that *practical reasons* are the source of irreversible information loss during the process of measurement. The root of this evil was discussed in Section 10.2: it is the additional noise that makes the simultaneous measurement of canonically conjugated variables possible. We might ask whether there is no other way to extract the full information about the system from measured data. Why do we have to perform simultaneous measurements? Is it not sufficient to measure different variables separately on parts of the ensemble of identically prepared systems? The unwanted penetration of vacuum fluctuations into the measurement apparatus would, in any case, be prevented.

It was realized early on that it is possible to reconstruct – though not always uniquely – the wave function from (separately measured) probability distributions for position and momentum; – however, an implementable experimental scheme is still lacking. It should also be pointed out that the same problem is known to appear in classical optics, namely in (optical and electron) microscopy. The role of the Schrödinger wave function is taken by the distribution of the complex classical field amplitude A in the object and the image plane, respectively. An intensity measurement immediately gives the distribution of $|A|^2$; however, the information about the field phase is missing. Its retrieval is the real physical problem. To this end, a further independent measurement is required. Such a measurement is the intensity measurement in the focal plane of the objective (exit lens). It is known that the Fourier transform of the field in the object plane is formed there, which corresponds to the change from the position to the momentum representation in quantum theory.

Let us concentrate on quantum optics and enquire about a practicable method of state reconstruction from measured data. For this purpose, we can exploit the amazing measurement possibilities offered by the homodyne technique. It was shown in Section 9.3 that, using this method, we can measure not only two chosen quadrature components of the field – the analogs of the position and the momentum – but a whole set of independent observables x_Θ (see Equation (9.12)). We emphasize that we are not dealing with simultaneous measurements; rather, we measure each variable on a part of the ensemble and determine its distribution. Indeed, a set of such distribution functions $\omega_\Theta(x_\Theta)$, with Θ varying in sufficiently small steps between 0 and π, contains the complete information about the quantum mechanical state of the system. This is generally valid; i.e. it applies not only to pure states but also to mixtures described by density matrices. It was shown by K. Vogel and H. Risken (1989) in a seminal paper that the Wigner function can be reconstructed from the distribution functions $\omega_\Theta(x_\Theta)$ via an integral transformation (see also Leonhardt and Paul (1995)). It is interesting to note that this transformation – it is the inverse radon transformation – has been known for a long time not only in mathematics; it is also the basis of medical computer tomography. Hence, the quantum mechanical reconstruction problem is, from the mathematical point of view, identical to the computer generation of an image of a body part (in the sense of an absorption profile) from a series of measured data which have been obtained by measuring the X-ray absorption of the object from different directions.

The discussed quantum mechanical measurements – now named optical homodyne tomography – have recently been realized experimentally (Smithey *et al.*, 1993). The Wigner functions of important quantum mechanical states of the light field, such as the Glauber state and the squeezed vacuum, were successfully reconstructed from measured data. As mentioned in Section 10.2, we obtain the density matrix from the Wigner function by a Fourier transformation, and because of this we can *calculate* the quantum mechanical expectation value for any desired variable and its probability distribution and declare the results as (indirectly) *measured*. In particular, "ideal" phase distributions can be so determined "experimentally."

Such a procedure is in contrast to the usual "measurement philosophy," which relies on the measurement of one or several observables, thus determining chosen experimental characteristics of the respective state, such as, for example, the photon statistics. Optical homodyne tomography, however, allows us to form a sort of holistic view on the quantum mechanical state. Having determined the Wigner function, we have everything at hand that is possible to know. The practicalities of this should not be underestimated: while the conventional method requires different measurement devices for different observables, optical homodyne tomography uses the same setup (in balanced homodyne tomography one needs only to change the phase of the local oscillator). In addition, optical mixing with a strong local oscillator allows us to work at considerably higher intensities, and hence we may use avalanche photodiodes (distinguished by a high detection sensitivity) for detection. It may be hoped that typically non-classical features, such as the comb-like structure of the photon distribution of the "squeezed vacuum," can be experimentally verified in this way, something that is not possible with conventional techniques.

Finally, let us comment on the latest developments related to quantum state reconstruction. It was proven theoretically, and then confirmed through extensive numerical simulations of measurements, that the density matrix of a quantum state (most favorably with respect to states with sharp photon numbers, i.e. in the Fock basis) can be reconstructed directly from the distribution functions $\omega_\Theta(x_\theta)$ (see Leonhardt 1997). The detour via the Wigner function was proven unnecessary, and we even improve the precision; i.e. we can extract finer details from the measured data. Also, it can be shown that less efficient detectors could be tolerated: it is possible, in principle, to reconstruct the density matrix, provided the detector efficiency is larger than $1/2$, but then we need a considerably increased measurement precision, which implies an increase in the number of measured data by orders of magnitude. This is in agreement with the statement made above that, in the process of smoothing the Wigner function – optical homodyne tomography with inefficient photodetectors enables us to reconstruct a smoothed Wigner function instead of the ideal one – the finer details are not lost *irretrievably*, it just requires much more effort to find the "truth."

11

Optical Einstein–Podolsky–Rosen experiments

11.1 Polarization entangled photon pairs

Throughout his life, Albert Einstein was never reconciled to quantum theory being an essentially indeterministic description of natural processes, even though he himself contributed fundamental ideas to its development. "God does not play dice" was his inner conviction. In his opinion, quantum theory was only makeshift. His doubts about the completeness of the quantum mechanical description were expressed concisely in a paper published jointly with Podolsky and Rosen (Einstein, Podolsky and Rosen, 1935). This paper analyzes a sophisticated Gedanken experiment, now famous as the Einstein–Podolsky–Rosen paradox, which has excited theoreticians ever since.

The Gedanken experiment was recently realized in a laboratory. The analyzed objects are photon pairs[1] – and this is what has motivated us to dedicate a chapter to this problem which has bearing upon the foundations of quantum mechanics. The photon pairs are formed by two photons generated in sequence (in a so-called cascade transition, as shown in Fig. 11.1). Due to the validity of the angular momentum conservation law (discussed in Section 6.9) for the elementary emission process, the two photons exhibit specifically quantum mechanical correlations, which are incompatible with the classical reality concept, as will be discussed in detail below.

How do the correlations appear in detail? Let us assume the initial state of the atom to be a state with angular momentum (spin) $J = 0$, the intermediate state to have angular momentum $J = 1$, and the final state to have again $J = 0$. The angular momentum conservation law implies for the system composed of the atom and the photons that the photon pairs must be in a state with total angular momentum zero. To satisfy this the two photons must "adjust" to one other, and we will analyze in detail what this means for their polarization states.

---

[1] In Einstein, *et al.* (1935), a system composed of two material particles prepared in a special state was considered.

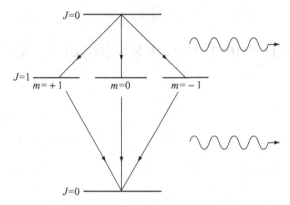

Fig. 11.1. Two-photon cascade transition. $J = $ atomic angular momentum; $m = $ magnetic quantum number.

The spatial separation of the sub-systems (in our case that of the photons) plays an important role in the Einstein–Podolsky–Rosen experiment. So, let us analyze such processes where the two photons have been emitted in opposite directions (we denote one direction as the x axis). Because the angular momentum $J = 1$ allows three orientations with respect to an arbitrarily chosen reference axis, the intermediate level of the cascade consists of three sub-levels with magnetic quantum numbers $m = -1, 0$ and $+ 1$. We choose the quantization axis orthogonal to the observation axis x, and denote it as the z axis. Corresponding to the three sub-levels, there are three transitions. The quantum mechanical calculation shows that the transition to the sub-level with $m = 0$ is associated with oscillations of the emitting electron in the z direction. On observation, we find that the radiation emitted in the x direction is linearly polarized in the z direction, as expected from the emission of a classical dipole. The transitions to the intermediate levels with $m = +1$ and $m = -1$ are each associated with a circular motion of the electron (one clockwise and the other anticlockwise) in the x, y plane. The part of the wave traveling in the x direction is therefore linearly polarized in the y direction.

The same motions of the electron take place in the transitions between the intermediate state and the final state of the cascade: when the transition starts from the level with $m = 0$, we are dealing again with oscillations in the z direction. The other two transitions are associated with circular motion in the x, y plane. The cascade process may take place via three channels, as shown in Table 11.1.

These three channels should not be understood in the sense of alternative routes. Only after a suitable measurement has been made can it be determined which channel was actually used. For example, if the first photon were found to be z

Table 11.1. *Channels of the cascade transition*
$J = 0 \rightarrow 1 \rightarrow 0$.

Channel	Intermediate level, magnetic number	Polarization direction	
		First photon	Second photon
1	$m = 0$	z	z
2	$m = +1$	y	y
3	$m = -1$	y	y

polarized, only the first channel is compatible with this result. We can see from Table 11.1 that the second photon will be polarized in the same direction as the first.

We could choose the z direction *arbitrarily*, apart from the constraint that it must be orthogonal to the observation direction. We can thus make the following general prediction. When a photon falls onto a detector with a polarizer in front of it and is detected, we can be sure that the other *unobserved* photon propagating in the opposite direction is polarized in the transmission direction of the polarizer.

Using a polarizing prism instead of a polarization filter, we can measure the polarization in two different directions. The prism splits the beam into two mutually orthogonally linearly polarized components which are, in addition, spatially separated. Performing a measurement on each of the partial beams with a separate detector for one incoming photon always results in one of the detectors responding and indicating one of the polarization directions (we assume ideal experimental conditions). This does not imply, however, that the detected photon was polarized in this way before it interacted with the measurement apparatus – rather the photon was transferred in the normal way into this state in the sense of a "reduction of a wave packet" by the measurement. We then know, based on the above information, that the second photon is in the same polarization state, and we do not need to perform another measurement.

The strong correlation between the polarization directions of the two photons listed in Table 11.1 implies that the spin part $|\psi_{tot}\rangle$ of the quantum mechanical wave function of the emitted field is a superposition of the states $|y\rangle_1|y\rangle_2$ and $|z\rangle_1|z\rangle_2$, where the state $|y\rangle_1$ describes the first photon linearly polarized in the y direction, etc. The relation between the superposition coefficients is determined by the requirement that the total wave function $|\psi_{tot}\rangle$ is the eigenfunction corresponding to the zero eigenvalue of the spin projection onto the propagation axis of the photons. (As we explained previously, in the cascade transition the total spin of the two photons – as a consequence of the angular momentum conservation law – must be zero, and this holds true in particular for the mentioned component.) The total wave function thus reads

$$|\psi_{\text{tot}}\rangle = \frac{1}{\sqrt{2}}(|y\rangle_1|y\rangle_2 + |z\rangle_1|z\rangle_2). \tag{11.1}$$

This is again a representative example of a quantum mechanically entangled state!

Linearly polarized light can also be viewed as a superposition of circularly polarized light, and so Equation (11.1) can be rewritten in the basis of circularly polarized states $|+\rangle$ and $|-\rangle$. Referring the sense of rotation (helicity) of the two photons to the same direction, say the x direction, we find the alternative representation

$$|\psi_{\text{tot}}\rangle = \frac{1}{\sqrt{2}}(|+\rangle_1|-\rangle_2 + |-\rangle_1|+\rangle_2), \tag{11.2}$$

and it is obvious that the spin in the x direction vanishes because for both terms the rotations of the field strength vectors compensate. From Equation (11.2) also follows the fact that when circular polarization detectors are used we find strong correlations: when one observer detects a left handed circularly polarized photon, the other observes the corresponding photon to be right handed circularly polarized, and vice versa. (Circularly polarized light can be measured with the help of the apparatus used for the detection of linearly polarized light with a quarter-wave plate placed in front of the polarizing prism.)

A discovery of great experimental importance was made in the mid 1990s. It was found that parametric fluorescence (Section 6.7) can be used for a very efficient generation of polarization entangled photon pairs. The interaction must be of type II because only in this case do two orthogonal polarization directions come into play. In the degenerate case (the signal and the idler wave frequency coincide) in a type I interaction, the propagation directions of the two photons lie on the same cone, while in a type II interaction they are on separate cones whose axes are tilted by an angle $\pm\theta$ with respect to the propagation direction of the pump wave. The polarization directions for the two cones are mutually orthogonal. Using a beta barium borate (BBO) crystal, and properly choosing the incidence angle of the pump beam with respect to the crystal optical axes (Kwiat *et al.*, 1995), the cones intersect along two straight lines (Fig. 11.2).

A photon propagating along such a line does not know to which cone it belongs, and therefore its polarization state is quantum mechanically uncertain. The other photon propagates along the other line and faces the same dilemma. We expect that the property of the two photons having orthogonal polarizations will not be

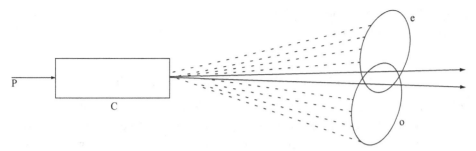

Fig. 11.2. Type II parametric fluorescence. The propagation directions of the two emitted photons each lie on a separate cone. P = pump wave; C = non-linear crystal; o = ordinary (vertical) polarization; e = extraordinary (horizontal) polarization.

lost under such special conditions. With respect to their polarization properties, the photons will be in a entangled state

$$|\psi_{\text{tot}}\rangle = \frac{1}{\sqrt{2}}(|x\rangle_1|y\rangle_2 + e^{i\alpha}|y\rangle_1|x\rangle_2). \tag{11.3}$$

The subscripts 1 and 2 refer to the propagation directions, and x (y) is the polarization direction of the ordinary (extraordinary) beam.

The phase α can be set to 0 or π with the help of a retarder (phase shifter from a birefringent material). The polarization direction can be, in addition, rotated by 45° by positioning a half-wave plate in one of the beams. In this way, any of the following four entangled states can be generated:

$$|\Psi^{\pm}\rangle = \frac{1}{\sqrt{2}}(|x\rangle_1|y\rangle_2 \pm |y\rangle_1|x\rangle_2), \tag{11.4}$$

$$|\Phi^{\pm}\rangle = \frac{1}{\sqrt{2}}(|x\rangle_1|x\rangle_2 \pm |y\rangle_1|y\rangle_2). \tag{11.5}$$

These form a complete basis of the Hilbert space of the polarization states of two photons. The states in Equations (11.4) and (11.5) are known as Bell states, and they play an important role in the quantum teleportation of an (arbitrary) polarization state (see Section 13.1).

The great advantage of this source of polarization entangled photon pairs is the fixed geometry of the two photon beams. The two photons can be easily selected using two pin holes (in principle, one would be enough as the photons are also strongly correlated in their propagation directions), or even better they can each be coupled directly into a glass fiber. The two-photon cascade does not have this property. When one photon is detected, the other photon will usually propagate anywhere other than towards the other detector. Therefore, parametric fluorescence of type II, as a generation mechanism of polarization correlated photons, is by far superior to cascade emission.

11.2 The Einstein–Podolsky–Rosen paradox

The quantum mechanical predictions based on the entanglement of the wave function – and this is the essence of the Einstein–Podolsky–Rosen paradox – are not compatible with the classical concept of reality. Indeed, the entangled states in Equations (11.1) to (11.5) must be interpreted in such a way that it is *uncertain in principle* what the polarization state of the photons is. It is of particular importance that the spatial separation of the photons can be made arbitrarily large. This ensures that any measurement performed on the first photon cannot physically affect the second photon in any way. According to the special theory of relativity, an action can travel at most with light speed; in the special case of the two-photon cascade, where the two photons travel in opposite directions, the action would chase the second photon without any chance of reaching it.

On the other hand, quantum mechanics assumes that the reduction is an instantaneous process. Then we can argue along with Einstein, Podolsky and Rosen in the following way. If a measurement is performed on the first photon, which indicates linear polarization, the other photon is, at that moment, definitely in a state of identical polarization (for the case of the wave function in Equation (11.1)). Because, as already discussed, the measurement cannot in any way influence the second photon, its polarization state could not have been changed, i.e. it had to be polarized in this way from the beginning.

However, we could have chosen to measure circular instead of linear polarization. In this case, we would come to the conclusion that – depending on the result of the measurement – the second photon is either left or right handed circularly polarized, and must have already been so before the measurement.

A *single* photon cannot be simultaneously linearly and circularly polarized, and we have arrived at the Einstein–Podolsky–Rosen paradox. What can we say now? We must state that there is no experimental contradiction. We are dealing with an unreal conditional clause of the form "If we had used another measurement apparatus, then we would have . . . ." Such a statement is not experimentally verifiable. We can in quantum theory – on a single system – perform only one type of measurement, and whatever another, unperformed, measurement would have yielded will remain speculation. In addition, we must realize that predictions of the form "This photon is linearly polarized" cannot be experimentally verified. We force the photon, through the choice of the measurement apparatus, to unveil itself as linearly or circularly polarized; i.e. we prescribe to the photon the type of the measured polarization state. The only thing that is left to the photon is the "decision" between the two orthogonal polarization directions or between left and right handed circular polarization. The above paradoxical statement "A single photon should be simultaneously linearly and circularly polarized" can be understood simply as that any photon can *arbitrarily* be detected as linearly or circularly polarized.

We have thus arrived at the root of the paradox. It does not make sense to ascribe definite properties (such as polarization) in the sense of the classical reality concept to a *single* microscopic system (for example, a photon). In other words, it is the non-objectifiability of the quantum mechanical description of nature which causes headaches; we are reminded once again that quantum theory is intrinsically a statistical theory which, in principle, does not allow a detailed description of single systems. It is difficult to avoid the impression that the quantum mechanical description is incomplete. A natural question arises as to whether the theory could be extended to remedy this flaw. Various scientists have hoped to reach this goal by introducing "hidden" (not accessible by measurement) variables.

The impossibility of such a dream was demonstrated rigorously in the Einstein–Podolsky–Rosen experiment. Owing to the importance of this point for the foundations of quantum mechanics, we will explain it in detail in Section 11.3.

11.3 Hidden variables theories

It is tempting to believe that uncertainty, which plays a central role in the quantum mechanical description of natural processes, is the consequence of our ignorance of very "fine" parameters. Knowledge of such parameters could allow us to predict, in *individual* cases, the results of arbitrary measurements, for example, at which moment a single radioactive atomic nucleus will decay. This knowledge lies at the heart of a (deterministic) hidden variables theory. Even though we cannot hope to ever obtain precise knowledge about the hidden variables, it is an interesting question for the theoreticians whether it is possible to believe in the existence of such a theory without penalties, i.e. without contradicting the predictions of standard quantum theory. In the positive case the theory would provide a deterministic "foundation" for quantum mechanics. Surprisingly, Bell (1964) could prove, using the example of the Einstein–Podolsky–Rosen experiment, that this is not so. Certain consequences of any hidden variables theory are not compatible with quantum mechanical predictions.

Let us explain the problem in some detail using the example of a realistic experiment, namely the measurement of the polarization properties of photon pairs generated in a cascade process – Bell himself studied Bohm's version of the Einstein–Podolsky–Rosen experiment involving two spin 1/2 particles prepared in the appropriate state. (Details can be found in the original paper by

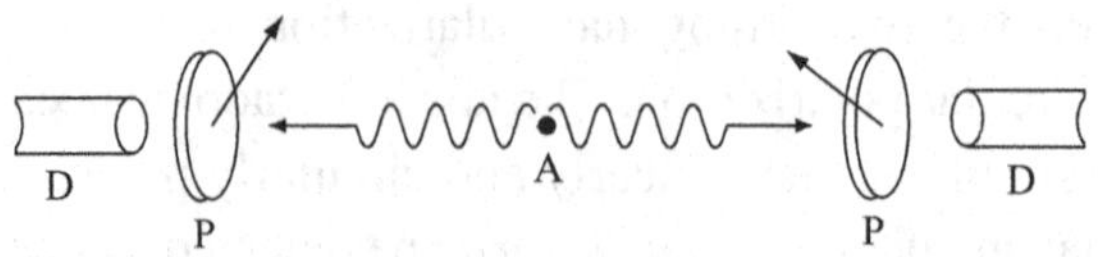

Fig. 11.3. Experimental setup for the measurement of coincidences on cascade two-photon transitions. A = atom; D = detector; P = polarizer; the arrows indicate the directions of transmittance of the polarizers. After Clauser *et al.* (1969).

Clauser *et al.* (1969) and also in the reviews by Clauser and Shimony (1978) and Paul (1980). Bell had the lucky idea to exploit the fact that we have a free parameter at our disposal in the experiment. We choose the experimental arrangement as shown in Fig. 11.3. The two photons propagating in opposite directions are detected by two detectors with polarizers placed in front of each of them.

We can set the orientations of the polarizers as we like, and, since the problem is rotationally symmetric, only the *relative orientation* of the transmission directions of the polarizers matters; hence, the angle Θ between the two orientations is, in this case, the above mentioned free parameter. The measured quantity is the coincidence counting rate $C(\Theta)$ of the two detectors observed for different values of Θ.

We describe the experiment using a deterministic hidden variables theory. In other words, we assume there exist hidden parameters which describe "in detail" the state of the radiation field after the cascade transition has taken place. The variables should be such that their particular values in any single case, together with the macroscopic parameters characterizing the setting of the measurement apparatus (in our case the orientations of the polarizers), allow the measurement results to be *unambiguously predicted*. It is of utmost importance for Bell's argument to adopt the reasoning of Einstein, Podolsky and Rosen explained in Section 11.2, which states that the measurement on one of the photons cannot influence the measurement on the other photon (due to the large spatial separation). We make, together with Bell, a locality assumption of the following form. Everything happening on the first detector is *exclusively* determined by the particular values of the hidden variables and the setting of the polarizer, and is completely independent of the orientation of the polarizer in front of the other detector.

With these assumptions, it is possible to show mathematically (with the limitation that the detectors are ideal, i.e. each incoming photon is detected) that the coincidence counting rate $C(\Theta)$ has to fulfil the following inequality (Freedman and Clauser, 1972):

$$-1 \leq 3\frac{C(\Theta)}{C_0} - \frac{C(3\Theta)}{C_0} - \frac{C_I + C_{II}}{C_0} \leq 0. \tag{11.6}$$

The value C_0 is the coincidence counting rate without polarizers inserted into the beam paths, and C_I and C_{II} are the coincidence counting rates with two polarizers.

The result in Equation (11.6) is extremely impressive when we recall the general assumptions under which it was derived. Because of this, we are not able to give a mathematical expression for $C(\Theta)$ itself. Nevertheless, we can say that $C(\Theta)$ must satisfy the limitations in Equation (11.6) whenever a deterministic theory which is, in addition, local in Bell's sense is used for the description.

The restriction in Equation (11.6) is drastic and, more importantly, it contradicts quantum mechanical predictions. This is easy to show. The quantum mechanical prediction is that, provided the first photon has been detected, the second photon is polarized in the transmission direction of the polarizer for the first photon (see Section 11.1). When the second photon impinges on a polarizer whose transmission direction is rotated with respect to the aforementioned direction by an angle Θ the following situation occurs. When we assume that the polarizers have 100% transmittivity the projection (in the classical description) of the electric field strength of the incoming wave onto the transmission direction of the polarizer is completely transmitted, i.e. only the fraction $\cos^2 \Theta$ of the intensity passes through. Because the photon is (as far as energy is concerned) indivisible, the detector behind the polarizer (which need not be ideal) will either detect the photon or not. The frequency of response (assuming the classical predictions to be still valid in the statistical mean) compared with the $\Theta = 0$ case decreases by a factor $\cos^2 \Theta$. The quantum mechanical coincidence counting rate as a function of the angle Θ thus takes the simple form

$$C^{\mathrm{qu}}(\Theta) = C^{\mathrm{qu}}(0) \cos^2 \Theta, \tag{11.7}$$

where the coincidence counting rate for equally oriented polarizers is denoted $C^{\mathrm{qu}}(0)$. Because, in this case, only half of the photons are detected, Equation (11.7) takes the form

$$C^{\mathrm{qu}}(\Theta) = \tfrac{1}{2} C_0^{\mathrm{qu}} \cos^2 \Theta, \tag{11.8}$$

where C_0^{qu} is the coincidence counting rate in the absence of both polarizers. The coincidence counting rate $C^{\mathrm{qu}}(0)$ remains unchanged when one of the polarizers is removed. (The second photon is, as we know, definitely polarized in the direction of the transmission direction of the polarizer, and hence is transmitted without problems.) Using the same notation as above, we have

$$C_{\mathrm{I}}^{\mathrm{qu}} = C_{\mathrm{II}}^{\mathrm{qu}} = C^{\mathrm{qu}}(0) = \tfrac{1}{2} C_0^{\mathrm{qu}}. \tag{11.9}$$

After inserting the quantum mechanical expressions from Equations (11.8) and (11.9) into Equation (11.6), we find that it is not satisfied for certain values of Θ.

The biggest discrepancies arise for $\Theta = \pi/8$ and $\Theta = 3\pi/8$. It is then advantageous to write down Equation (11.6) for these two angles and then to subtract the two relations. We thus obtain the following simple inequality:

$$|\frac{C(\frac{\pi}{8})}{C_0} - \frac{C(\frac{3\pi}{8})}{C_0}| \leq \tfrac{1}{4}, \qquad (11.10)$$

which has the advantage that the coincidence counting rates C_I and C_{II} no longer appear. It is easily checked that the quantum mechanical formula in Equation (11.8) is in contradiction with Equation (11.10); it yields the value $\sqrt{2}/4 \approx 0.35$ on the left hand side.

Hence, we must bury our hopes that it is possible to alter quantum mechanics without touching its quantitative predictions, through the introduction of hidden variables, and so put it onto a "sound", i.e. classical-deterministic, basis. There are – under the conditions existing in a Einstein–Podolsky–Rosen experiment – correlations between two sub-systems which defy classical understanding and are therefore of a purely quantum mechanical nature.

Let us illustrate this once again using the example of the two-photon cascade. Let us consider the setup depicted in Fig. 11.3, but now the polarization is measured on each beam in two orthogonal directions e and f. This can be achieved with the help of a polarizing prism and *two* detectors positioned behind it. Assuming both prisms to be equally oriented, we find striking correlations between the measured data; the two measurement apparatuses indicate the same polarization state: the two photons (forming one pair) prove to be oriented either both in the e direction or both in the f direction, and these two cases appear, without statistical irregularities, with equal probability.

This fact, were one to describe it on the basis of the classical concept of reality, could be understood only under the assumption that the photons had the given (in the measurement, found) polarization right from the beginning, i.e after being emitted. The source could not emit photon pairs polarized in a different direction g ($\neq e$ or f) because then there will unavoidably occur events in which both detectors register different polarizations. (We assume from experience that a photon polarized in the g direction will be detected, with a non-zero probability, as if it were polarized in the e direction, and as if it were polarized in the f direction, with a different, but also non-zero probability.) This picture is incompatible with the fact that the discussed correlations – due to the rotational symmetry of the problem with respect to the line connecting both detectors – fully persist when both crystals are rotated by the same angle.

The failure of all attempts to objectify the polarization properties of the emitted photons comes as no surprise when it is viewed from the quantum mechanical formalism. As discussed at length in Section 11.1, we know that both photons are in an entangled quantum mechanical state, and this means that the polarization of the two photons must be uncertain in the *quantum mechanical* (not interpretable as simple ignorance) sense.

A measurement of the polarization on one of the photons leads, according to the quantum mechanical rules for the description of the measurement process, to a reduction of the wave packet, and the unobserved second photon (depending on the measurement result) is brought into one or the other polarization state. According to the two possible representations, Equations (11.1) and (11.2), of the photon wave function, we can polarize the second photon linearly or circularly at will. Because the only constraint for the x direction is that it must be orthogonal to the propagation direction of the photons, we can choose in addition the orientation of the "crossed axes" of the linear polarization by rotating the measurement apparatus around the propagation direction. The reduction is instantaneous, as assured by quantum mechanics, and hence it seems we face a faster than light transfer of action onto the second photon such that the uncertainty present before the measurement is removed. Whatever *physical* process is hidden behind the reduction, one thing can be taken as granted: it is basically impossible to detect a physical effect (on the individual sub-system). What is observable is the *change* of a physical variable, and this requires well defined initial values. This is, however, not the case in the present situation, and hence the reduction of the wave function cannot be used for signal transmission. We return to this point for a more detailed discussion in Section 11.5.

First, let us analyze an analogous experiment in classical optics. An unpolarized light beam is split, with the help of a polarization-insensitive beamsplitter, into two (also unpolarized) beams, which propagate in different directions. We have to picture unpolarized light classically as follows. When we measure, within any coherence time interval, a certain (random) polarization (in general, elliptical), we detect a different polarization within the next time interval. Observing the light over a long time interval, we find all possible polarization states with equal probability. When an observer measures the instantaneous polarization state in one of the partial beams (classically this is not a problem), the polarization state of the other light beam at the same distance from the beamsplitter is revealed. It is exactly the same because the beamsplitter does not change the (instantaneous) polarization. The information gain is, in this case, also instantaneous. Naturally, there is no physical influence on the unobserved sub-system from the measurement. We only find out a fact that already existed objectively though it was unknown because of the statistical changes of the polarization state. This is the essential difference between classical and quantum theory.

We have mentioned several times that quantum mechanical predictions can be verified only on ensembles. So, let us look at the ensemble of photon pairs! As above, let us assume that both observers oriented their linear polarization measurement apparatuses in the same way. Each observer does not notice actions taken by his colleague. The measurement apparatus of each observer indicates that the photons are randomly polarized – with equal probability – in the e and f directions. Actually, the observers are dealing with unpolarized light! Only after they compare their measurement results do they realize (as explained) that they agree completely. The same also happens when circular polarization is measured (we, as usual, relate the sense of rotation to the propagation direction).

We can also think of a realization of the experiment where the first observer makes a measurement earlier than the second one (Paul, 1985). Then the first observer is enabled to play the role of a "clairvoyant." Let us imagine that the second beam, after traversing a certain distance, is diverted by a mirror such that it comes close to the first beam. The two observers can then position themselves next to one another. The first observer can now predict the measurement results as there is a time delay before the second photon arrives (we assume that the source emits photon pairs one after the other). The second measurement is thus unnecessary. The second observer can use the obtained information to divide the ensemble into two parts with well defined polarization properties; in the present case, into photons linearly polarized in the f and e directions. The statement that we have sub-ensembles with sharp polarization can be experimentally verified: a corresponding measurement on each member of the respective ensemble will give the same result.

The first observer could have rotated the apparatus to measure the polarization with respect to two other directions, say e' and f'. A randomly selected photon from the second beam, which belongs in the case of the first measurement to the ensemble of, say, e polarized photons, would neatly fit into a sub-ensemble with another polarization direction (e' or f'). The first observer could have used, in addition, a measurement apparatus for circular instead of linear polarization. A photon from the second beam would then have become a member of a group of either only left or right handed circularly polarized photons.

A single photon is thus unmasked as a pure "opportunist", adjusting without any problems to different situations, and there are no "independent values" such as real polarization properties.

The described experiment can also be interpreted in such a way that the second beam, which is by itself unpolarized, can be split in different ways into two sub-ensembles of orthogonal polarizations. That such decomposition is *thinkable* is easily seen from the form of the density matrix. The experiment proves that this is indeed also feasible. The decomposition of a quantum mechanical mixture into different "components" might seem surprising at first glance. This is, however, a general peculiarity of the quantum mechanical description. For instance, there is experimentally no detectable difference between two ensembles formed in the following way: the sub-ensembles are in Glauber states with equal amplitude and phases uniformly distributed between 0 and 2π, or they are in states with sharp photon numbers (Fock states) with a Poisson weight function. The density matrix is exactly the same in both cases.

Finally, let us comment on the quantum mechanical description of a *single system* using a wave function. This means moving on to shaky ground! Following the basic principles of quantum theory, the wave function is always related to a whole ensemble of systems. In particular, the quantum mechanical probability predictions can be verified only on an ensemble. We pointed out in Section 11.2 that it does not make sense to ascribe objective properties to a single system. Nevertheless, the wave function provides us with valuable experimental hints. Knowing, for example, that a photon is in the polarization state $|y\rangle$, we can predict with certainty that the photon will pass a y oriented polarization filter (under ideal conditions) but will not pass an orthogonally oriented filter. An important question is how do we obtain the information required about the single system. We need to know the preparation process in detail. For example, we have "fabricated" the polarized photon from a weak unpolarized beam that was made to impinge on a polarization filter. However, it is sufficient that we watch the preparation, or gain the information from the person conducting the experiment (who is assumed to be trustworthy). Another possible preparation method was opened up by the Einstein–Podolsky–Rosen experiments. The situation is completely different when we are asked to determine the polarization state of any given single photon. In such a case, our position is hopeless; we saw in Section 10.3 that state reconstruction can be realized only on an ensemble. We are in a strange position: if we are lucky we know the state of a single system, but we cannot determine it if we are unaware of the process of preparation.

Let us now return to the Einstein–Podolsky–Rosen experiment. The most astonishing property of the observable correlations is that they are not limited to microscopic dimensions, but actually extend over macroscopic distances. Are the quantum mechanical predictions trustworthy under such extreme conditions? This question seems completely justified, and Schrödinger (1935) was the first sceptic. He considered it possible that the correlations between the parts of a system break down spontaneously when their separation exceeds a critical value. For photons, such a distance can only be the coherence length. Indeed, Bell's locality assumption is physically justified only when the photons belonging to one pair – conceived as wave trains whose length in the propagation direction is given just by the coherence length – are "liberated" from the atom. The experimental examination of equation (11.10) recently performed by several authors is thus of interest, not only from the viewpoint of making an experimental decision for or against hidden variables theories, but represents at the same time a test of quantum mechanics under extraordinary conditions. The following section analyzes briefly the experimental situation, but let us state the result beforehand – quantum mechanics was perfectly verified.

11.4 Experimental results

Before we say a few words about real Einstein–Podolsky–Rosen experiments in the form of coincidence measurements discussed in Section 11.3, we will comment on a fundamental difficulty in Bell's argument caused by realistic, i.e. low efficiency detectors. The derivation of Equation (11.6) relies to a large extent on the assumption that all incoming photons are detected. When we drop this assumption, the inequalities obtained have no physical relevance. This motivated Clauser *et al.* (1969) to modify the assumptions about the action of the hidden variables: these should (together with the macroscopic parameters) determine the passing or not passing of the photon through the polarizer rather than the action of the detector. With this concept, the inequalities in Equations (11.6) and (11.10) still hold.

The first experiment was performed by Freedman and Clauser (1972). To guarantee an undisturbed two-step emission process (assumed in the theoretical treatment), they used as a light source a beam consisting of Ca atoms excited to the starting level of the cascade transition by resonant absorption of radiation from a deuterium arc lamp. The analysis was rather difficult because the coincidence counting rate contained a large contribution of random coincidences. We have to keep in mind that the atoms rarely emit the two photons in opposite directions. When one of the photons of the pair is detected, its partner will, in most cases, be flying somewhere else in a different direction. It often happens that the detectors register photons belonging to different pairs because many atoms emitted simultaneously. Such coincidences are obviously purely random.

Freedman and Clauser determined the random coincidence counting rate separately by measuring coincidences with large time delays. They determined the systematic coincidence counting rate to which the theoretical predictions refer by subtracting the random coincidence counting rate from the undelayed counting rate. Because, even in the absence of polarizers, only about 0.2 coincidences per second were counted, it took at least 200 hours to obtain statistically reliable results.

Freedman and Clauser found a significant violation of the inequality in Equation (11.10); on the other hand, the quantum mechanical predictions were in excellent agreement with the measured data. Later experiments, in which the use of electron beams or laser light helped to excite atoms in much greater numbers, drastically shortening the measurement time, led to similar results (for details, see Clauser and Shimony, 1978 and Paul, 1980. A common feature of all these experiments was the short distance between the detectors (to avoid intensity losses). A separation of the detectors by a distance much larger than the coherence length of the investigated radiation was realized for the first time in the experiment by Aspect, Grangier and Roger (1981). They demonstrated not only that the measured data verified the quantum mechanical prediction for all angles between the transmission directions of the two polarizers with unprecedented statistical precision, but also that this data did not change when the distance of the polarizers from the light source was made as large as 6.5 m. Aspect *et al.* hence satisfied an important condition underlying Bell's locality assumption.

There is one further requirement. The locality postulate is necessary only when it is demanded by the causality principle. This implies that the orientation of at least one of the two apparatuses must be changed quickly enough during the measurement to ensure that the information about the actual orientation of one apparatus cannot reach the other while the measurement takes place there. Causality enters the situation at this point: a signal cannot be transmitted faster than light (orienting the polarizers before the beginning of a measurement sequence in a certain way would give the measurement apparatuses enough time to exchange information). A later experiment by Aspect, Dalibard and Roger (1982) also incorporated fast switching of both measurement apparatuses. These authors used a technique similar to that applied in the "delayed choice" experiment described in Section 7.3. Both photons had the choice of two possible paths to reach the detection apparatuses which were equipped with differently oriented polarizers. An acousto-optical switch "decided" which way was open. The switch consists of a liquid cell with two opposite in-phase driven electro-acoustic transducers, which generate a standing ultrasound wave in the liquid; i.e. a density variation acts as a diffraction grating (called the Debye–Sears effect). The incoming wave is completely deflected to the side in a fixed direction for a high enough amplitude of the ultrasound wave; however, it leaves the liquid unaffected when the amplitude is zero. In this way the incoming beam is periodically deflected. To come closer to the ideal random switching, Aspect *et al.* (1982) used two high frequency generators of different frequencies to drive the two acousto-optical modulators (one for each photon). This sophisticated experimental setup led straight to the statement that quantum mechanics is, as always, right.

The experimental results are in distinct contradiction with Equation (11.10); i.e. they are incompatible with any, however refined, (deterministic) hidden variables theory satisfying Bell's locality postulate. This is a result Einstein probably least expected when, in 1935, he (with Podolsky and Rosen), with his profound criticism, started the development leading to the described impressive theoretical and experimental accomplishments.

Recently, the construction of photon pair sources using parametric interaction of type II underwent noteworthy progress. In an experiment performed by the Zeilinger group (Weihs *et al.*, 1998), the photons of a pair were each coupled into a single-mode optical fiber. The receiving stations were at a distance of 400 m. In addition, the "last minute" procedure of polarization orientation of Aspect *et al.* (1982) was perfected. Scrupulous critics could still argue that the acousto-optical modulators switch not randomly but deterministically. This would give the photons, in principle, the opportunity to foresee which polarizer setting they will find. Aspect *et al.* closed this loop-hole by letting quantum mechanical randomness rule absolutely. They used a beamsplitter illuminated by a light emitting diode, with a detector at each exit, as a random number generator. The light intensity was kept very low to avoid coincidences (when they happened, they were ignored). The response of the one or the other detector was the desired random process delivering digital zeros and ones. In the latter case the signal was used after amplification to drive an electro-optical modulator. It caused a rotation of the polarization direction of the photon by an angle proportional to the applied voltage. The photon impinged again on a polarizer, or, more precisely, a polarizing prism with a detector at each of the exits. Instead of rotating the polarizing prism, the polarization direction of the photon was rotated in a definite way, the response probabilities of the detectors being the same in both cases. Also in this experiment a significant violation of Bell's inequality (in a slightly more general form than in Equation (11.6)) was observed, and agreement with the quantum mechanical prediction was excellent. A particular advantage of the novel source of photon pairs is the fixed propagation directions of the polarization correlated photons, so we are spared the abundance of random coincidences known from the two-photon cascade.

This experiment is especially impressive because the Einstein–Podolsky–Rosen correlations extend over a macroscopic distance (400 m). Further, using the glass fiber net used by the Swiss communications company Swisscom (Tittel *et al.*, 1998) quantum correlations at a distance of 10 km were observed, for photons entangled in energy and time (see Section 11.6). There are no signs of a spontaneous collapse of such correlations for large spatial separations of the subsystems. If there were such a collapse, for instance in the case of the two-photon cascade, the foundations of our description of nature would break down. The disappearance of quantum mechanical correlations would contradict the angular momentum conservation law: the initial sharp value (zero) of the angular momentum of the two-photon system would become necessarily non-sharp by such a process.

From this point of view, a theoretician will accept the results of the experiments with satisfaction.

Finally, we would like to point out that the great range of the quantum mechanical correlations characteristic of the Einstein–Podolsky–Rosen experiment can be deduced from simple optical beam-splitting (for details, see Paul (1981)). For initial sharp photon number states split by the beamsplitter, we are actually dealing with an experiment of the abovementioned type. On the one hand it is possible, by measuring the photon number at one of the outputs (the reflected or the transmitted), to predict with certainty how many photons will be detected at the other output. On the other hand, it is known that the two partial beams can be made to interfere, which means that the relative phase must be fixed. By measuring the phase of one of the beams, we obtain precise information about the phase of the other beam. We can now repeat the argument of Einstein, Podolsky and Rosen given in Section 11.2, and we come to the following conclusion. Imagine that we separate sufficiently the partial beams: the photon number, as well as the phase, should be well defined in each of the partial beams from the beginning; this, however, is not possible according to quantum mechanics. The disappearance of the quantum mechanical correlations between the two beams would destroy their ability to interfere, but such loss of interference has not been observed even for considerable spatial separation (before their reunion) – let us recall the experiment of Jánossy and Náray (1958) with an interferometer arm length of 14.5 m – nor would such an occurrence be expected by any theoretician to occur, however large the separation might be.

11.5 Faster-than-light information transmission?

There has been daring speculation about the possibility of using the Einstein–Podolsky–Rosen correlations for information transmission, leading to the possible exploitation of the instantaneous character of the reduction of the wave function – caused by a measurement performed on one of the sub-systems – for faster-than-light information transmission. This would imply a dramatic violation of causality, and hence we *could* dismiss the whole idea as bizarre. However, a detailed analysis why this does not work helps us to learn a great deal about the physics involved, and so we include it here.

What should such an experiment look like? Let us think of two observers (far away from each other) performing polarization measurements on the photons belonging to a photon pair emitted in a cascade transition (Fig. 11.3). One of them, say observer A, wishes to send a message to observer B. The first consideration is that of information coding. The use of the polarization states of the photons is certainly a good choice. An important feature of this experiment is the randomness of the individual polarization state – when, for instance, observer A chooses the setup for measuring linear polarization along the x and the y direction he has no influence on whether a randomly selected photon will "decide" in favor of the x or y direction. This rules out a simple coding such as x polarization means zero, y polarization means one. However, it seems that the difference between circular and linear polarization could be exploited; for example, observer A could perform the coding by switching at will between the two measurement setups. As mentioned in Section 11.1, this could be accomplished by inserting a quarter-wave plate into the optical path, or removing it, respectively. The arrangement would be, for instance, that circular polarization means zero and linear polarization means one. Because the measurement changes the polarization state of the other photon *instantaneously* into the state of the measured one (linear or circular polarization) – at least quantum theory asserts this (Section 11.2) – the task left for observer B is (after a short time delay necessary to ensure that the reduction has taken place) to identify the polarization (circular or linear) of the photon, and thus to obtain the information. However, this is not possible! It was shown in detail in Section 11.2 that the polarization state (circular or linear) of a single photon is prescribed by the measurement apparatus. It is absolutely impossible to deduce from a single measurement the polarization state of the photon before the measurement. It makes no physical sense to ascribe to a single photon a polarization property in the sense of an objective characteristic. We come to the interesting conclusion that the non-objectifiability of the quantum mechanical description (which is usually not taken too seriously) is the price paid for upholding causality, and is thus of fundamental importance. The discussion initiated by Einstein, Podolsky and Rosen has an additional merit: it disclosed an unexpected and close relationship between the non-objectifiability in the micro-cosmos and causality – the fundamental principle of (mainly) macroscopic physics.

We can draw some additional conclusions. A sceptic might ask the following questions. Do we have to perform a direct measurement? Couldn't we somehow make a lot of copies of the photon first and then measure the polarization (this would not cause any problems as we have a whole ensemble of identical particles at our disposal)? Taking the reasonable standpoint that a faster-than-light signal transmission is impossible, since otherwise the whole foundations of physics would crumble – we could influence the past, and, to give an example, I could kill my father before he procreated me – we can say beforehand that copying *cannot* be successful. It implies in particular that it is not possible to "clone" an individual photon, i.e. to produce one copy and hence an arbitrary number of *identical copies*. This no go theorem can also be proven directly. It could be shown that a possible "cloning" is in direct contradiction to the linearity of the Schrödinger equation (Wootters and Zurek, 1982). This result is known as the "no cloning theorem."

Another consequence is that there is no amplifier which would allow us to "read out" from the amplified field the polarization state (linear or circular) of the original photon. It is not obvious that this should be so. Certainly it is not possible to avoid amplifier noise which distorts the amplified field, but should it not make a difference, even when amplifying a single photon, whether the initial signal was linearly or circularly polarized? Indeed, performing a quantum mechanical calculation for the initial condition "we start with a photon with a well defined polarization," we naturally find that the light leaving the amplifier depends on the polarization of the initial photon. However, we realize again the serious limitation imposed on the quantum mechanical description by the ensemble interpretation: we find a "trace" of the original polarization in the amplified field (and from the measured data it is possible to infer the polarization), but this holds only when we investigate the whole ensemble, i.e. when we repeat the experiment under exactly the same conditions many times. It should be stressed yet again that quantum mechanics does not make any detailed predictions about a single experiment.

A detailed theoretical analysis of an (idealized) polarization insensitive amplification process leads to the following results (Glauber, 1986): in each single case, the amplified field is in a well defined polarization state. Independently of the initial photon polarization, *any* polarization state (i.e. elliptic polarization) is possible. Differences are found when we ask for the probability of finding a given polarization of the amplified field in dependence on the polarization of the original photon. As is to be expected, the probability is largest when both polarizations coincide. However, the probability is not zero (due to amplification noise), even when the two polarization states are orthogonal. Only by measuring the probability distribution – which can be determined only on an ensemble of amplified fields – can the polarization of the identical original photons be determined. (The measurement of the polarization state itself is, for sufficiently large amplification, not a problem. In such a case we are dealing with classical fields, so standard methods of classical polarization optics are applicable.) In contrast, a single measurement does not give any information about the polarization of the initial photon.

However, as mentioned previously, our goal is only to distinguish between linear and circular polarization of the initial photon. We can give a simple argument when we work with larger redundancy. By this we mean the following: observer A, having the desire to communicate at a speed faster than light, encodes the information not into single photons but each bit into an ensemble of many photons forming a whole sequence. Accordingly observer A does not change the setup during time intervals of duration Δt. Let us imagine the sender to be far away from us; i.e. the time Δt is still much shorter than the time light needs to reach us (there is no fundamental limitation for the distance). The aforementioned ensembles are formed from differently polarized photons depending on whether they are to communicate a zero or a one. In one case the photons are linearly polarized in the x or the y direction, and in the other case they are left or right handed circularly polarized. We saw in Section 11.3 that such ensembles are in fact physically identical: they describe unpolarized light. We have not the slightest chance of distinguishing between them experimentally, whatever clever technique we might employ; and an amplifier will be of no help either, nor can it save the faster-than-light information transmission (even though researchers in this field had believed, or at least hoped, it would).

11.6 The Franson experiment

In Section 6.7 we mentioned two important features of photon pairs generated by parametric fluorescence, namely the simultaneity of the two-photon emission and the frequency relation $\omega_p = \omega_s + \omega_i$ (see Equation (6.8)). Because both photons have broad bandwidths $\Delta\omega_s$ and $\Delta\omega_i$, the frequencies ω_s and ω_i have to be understood as frequencies within the corresponding bandwidths, as they appear in the measurement on a single photon pair. It is the energy conservation law, which is also valid for individual systems, that is behind this frequency relation, as explained in Section 6.7. The "instance of emission" is, however, not determined. The non-linear crystal is pumped continuously and the emission is spontaneous. As in the case of spontaneous emission from an excited atom, the emission moment is not predictable.

The American scientist Jim Franson was inspired by these unusual properties of the photon pairs, often referred to as energy and time entangled, and he proposed an original experiment,[2] which will be discussed in some detail below (Franson, 1989). The idea was to make coincidence measurements of photon pairs (signal and idler photons fall onto separate detectors positioned at the same distance from the source, and only events with simultaneous detector response are registered) more exciting by placing in front of each detector one of two identical Mach–Zehnder interferometers (see Fig. 11.4).

The transit time difference in the interferometer, ΔT, should be large compared with the inverse of the bandwidths of the signal and the idler wave. Representing the photons in a classical picture by coherent light pulses, the condition requires the duration of the pulses to be short compared with ΔT. This implies that the two partial pulses originating from a photon impinging on the interferometer (one of the pulses carries straight on, the other takes a detour prescribed by the interferometer) never meet again, in particular not on the detector. There is no place for the "interference of the photon with itself." However, interference effects show up

---

[2] Actually, Franson considered photon pairs from a cascade transition, which are also energy and time entangled.

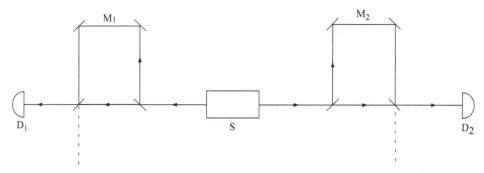

Fig. 11.4. The Franson experiment. Coincidences between detectors D_1 and D_2 are registered. S = source of entangled photon pairs; M_1, M_2 = identical Mach–Zehnder interferometers.

in the coincidence counting rate (noticed by Franson, to his credit), whereby the mentioned frequency condition plays a decisive role.

We will try to understand the effect with the help of classical optics. Because the response probability of a detector is proportional to the instantaneous intensity $I(\mathbf{r}, t)$ residing on its sensitive surface, the coincidence frequency reads, in this case (up to a constant factor),

$$W_c = I(\mathbf{r}_1, t) I(\mathbf{r}_2, t) = E^{(-)}(\mathbf{r}_1, t) E^{(+)}(\mathbf{r}_1, t) E^{(-)}(\mathbf{r}_2, t) E^{(+)}(\mathbf{r}_2, t)$$
$$= |E^{(+)}(\mathbf{r}_1, t) E^{(+)}(\mathbf{r}_2, t)|^2. \tag{11.11}$$

The positions of the two detectors are $\mathbf{r}_1$ and $\mathbf{r}_2$, and we have expressed the intensities according to Equation (3.10) through the positive and negative frequency parts of the electric field strength.

The presence of the interferometers leads to the following expressions for the variables $E^{(+)}$ on the detector surfaces:

$$E^{(+)}(\mathbf{r_1}, t) = \tfrac{1}{2} \left[E_s^{(+)}(\mathbf{r_1}, t) + E_s^{(+)}(\mathbf{r_1}, t - \Delta T) \right], \tag{11.12}$$

$$E^{(+)}(\mathbf{r_2}, t) = \tfrac{1}{2} \left[E_i^{(+)}(\mathbf{r_2}, t) + E_i^{(+)}(\mathbf{r_2}, t - \Delta T) \right]. \tag{11.13}$$

In writing these equations we have assumed that the signal wave hits the detector at position $\mathbf{r_1}$ and that the idler wave strikes at position $\mathbf{r_2}$. We expressed the superposed electric fields through the field strengths found in the absence of the interferometers. Equations (11.12) and (11.13) express the fact that the part of the pulse taking the direct route meets the part of the pulse making the detour (in the interferometer), and hence delayed by ΔT, at the detector surface. The factor of $1/2$ originates from the splitting of the electric field strength on the beamsplitter; the transmitted and reflected parts are attenuated by a factor of $1/\sqrt{2}$.

The term relevant for the coincidences according to Equation (11.11), $E^{(+)}(\mathbf{r}_1, t)E^{(+)}(\mathbf{r}_2, t)$, takes the following form:

$$E^{(+)}(\mathbf{r}_1, t)E^{(+)}(\mathbf{r}_2, t) = \frac{1}{4}\left[E_s^{(+)}(\mathbf{r}_1, t)E_i^{(+)}(\mathbf{r}_2, t) + E_s^{(+)}(\mathbf{r}_1, t - \Delta T) \right.$$
$$\times E_i^{(+)}(\mathbf{r}_2, t - \Delta T) + E_s^{(+)}(\mathbf{r}_1, t)E_i^{(+)}(\mathbf{r}_2, t - \Delta T)$$
$$\left. + E_s^{(+)}(\mathbf{r}_1, t - \Delta T)E_i^{(+)}(\mathbf{r}_2, t) \right]. \tag{11.14}$$

First, let us consider the third and fourth terms on the right hand side of Equation (11.14). The squared absolute value of the third term, for instance, determines the probability that the detector at position $\mathbf{r}_1$ registers a (signal) photon at time t, and the other registers an (idler) photon at time $t - \Delta T$. Because the emission of the photons is simultaneous and their extension in time is much smaller than ΔT, the probability must be zero. The third term is therefore negligible, and the same holds for the fourth term.

The first term on the right hand side of Equation (11.14) describes a coincidence at time t in the absence of the interferometers. Assuming the pulses to be classical wave packets, we can rewrite the first term as

$$E_s^{(+)}(\mathbf{r}_1, t)E_i^{(+)}(\mathbf{r}_2, t) = \sum_{\mathbf{k}_s, \mathbf{k}_i} c_{\mathbf{k}_s}^{(s)} e^{i(\mathbf{k}_s\mathbf{r}_1 - \omega_s t)} c_{\mathbf{k}_i}^{(i)} e^{i(\mathbf{k}_i\mathbf{r}_2 - \omega_i t)}. \tag{11.15}$$

Here the complex amplitudes determining the pulse form are denoted by $c_{\mathbf{k}_s}^{(s)}$ and $c_{\mathbf{k}_i}^{(i)}$. The summation runs over all wave number vectors $\mathbf{k}_s$ and $\mathbf{k}_i$ present in the pulse (more precisely, integrations should be taken instead of the summations, but for our argument this does not matter). The frequencies ω_s (ω_i) are those related to the wave number vectors $\mathbf{k}_s$ ($\mathbf{k}_i$).

The really interesting term in Equation (11.14) is the second one. According to Equation (11.15) it can be expressed as

$$E_s^{(+)}(\mathbf{r}_1, t - \Delta T)E_i^{(+)}(\mathbf{r}_2, t - \Delta T)$$
$$= \sum_{\mathbf{k}_s, \mathbf{k}_i} e^{i\omega_s \Delta T} c_{\mathbf{k}_s}^{(s)} e^{i(\mathbf{k}_s\mathbf{r}_1 - \omega_s t)} e^{i\omega_i \Delta T} c_{\mathbf{k}_i}^{(i)} e^{i(\mathbf{k}_i\mathbf{r}_2 - \omega_i t)}. \tag{11.16}$$

It is easy to see that the two sums virtually vanish when they take their maximum or a value close to it, for $\Delta T = 0$. The reason for this is the inclusion of factors $e^{i\omega_s \Delta T}$ and $e^{i\omega_i \Delta T}$ (the products $\Delta\omega_s \Delta T$ and $\Delta\omega_i \Delta T$ are large compared with unity by assumption), which undergo several oscillations in the summation. The only non-vanishing term in Equation (11.14) is, therefore the first one, and this implies, according to Equation (11.11), that the probability of counting a coincidence is independent of the time delay ΔT caused by the interferometer; i.e. there are no interference effects.

The obtained result is not too surprising as we tacitly assumed the signal and idler pulses not to be correlated, as can be seen from the factorized form in Equation (11.15). It is a straightforward matter to amend this relation to account for correlations. It is necessary to replace the product $c_{\mathbf{k}_s}^{(s)} c_{\mathbf{k}_i}^{(i)}$ by a new coefficient $c_{\mathbf{k}_s, \mathbf{k}_i}$. Because both waves are generated by parametric fluorescence, we will require this coefficient to be non-zero only when the phase matching condition Equation (6.7), as well as the frequency condition, Equation (6.8), are satisfied. This rather pragmatic attitude leads to a surprising result. The double sum replacing Equation (11.16) is no longer affected by exponential factors depending on ΔT, since they can be extracted as a common prefactor $e^{i\omega_p \Delta T}$. The results can be summarized as follows.

Using the notation

$$\Psi(\mathbf{r}_1, \mathbf{r}_2; t) = \sum_{\mathbf{k}_s, \mathbf{k}_i} c_{\mathbf{k}_s, \mathbf{k}_i} e^{i(\mathbf{k}_s \mathbf{r}_1 - \omega_s t)} e^{i(\mathbf{k}_i \mathbf{r}_2 - \omega_i t)}, \qquad (11.17)$$

the coincidence probability can be written, according to Equations (11.11) and (11.14), in the following form:

$$\begin{aligned} W_c &= \tfrac{1}{16} |\Psi(\mathbf{r}_1, \mathbf{r}_2; t) + \Psi(\mathbf{r}_1, \mathbf{r}_2; t - \Delta T)|^2 \\ &= \tfrac{1}{16} |\Psi(\mathbf{r}_1, \mathbf{r}_2; t) + e^{i\omega_p \Delta T} \Psi(\mathbf{r}_1, \mathbf{r}_2; t)|^2 \\ &= \tfrac{1}{16} |\Psi(\mathbf{r}_1, \mathbf{r}_2; t)|^2 |1 + e^{i\omega_p \Delta T}|^2. \end{aligned} \qquad (11.18)$$

This relation is easily generalized to the case when an additional optical element, making the phase of the light larger by Φ_s or Φ_i, respectively, is inserted into the longer arms of the interferometers. As a consequence, the positive frequency part of the electric field strength making the detour must be multiplied by the phase factor $\exp(-i\Phi_s)$ or $\exp(-i\Phi_i)$, respectively. Equation (11.18) thus becomes

$$\begin{aligned} W_c &= \tfrac{1}{16} |\Psi(\mathbf{r}_1, \mathbf{r}_2; t)|^2 |1 + e^{i(\omega_p \Delta T - \Phi_s - \Phi_i)}|^2 \\ &= \tfrac{1}{8} |\Psi(\mathbf{r}_1, \mathbf{r}_2; t)|^2 [1 + \cos(\omega_p \Delta T - \Phi_s - \Phi_i)]. \end{aligned} \qquad (11.19)$$

The results obtained are, in fact, in full agreement with quantum mechanical prediction (Franson, 1989).

Equation (11.19) allows the following simple interpretation: there is a probability amplitude $\Psi(\mathbf{r}_1, \mathbf{r}_2; t)$ for measuring a coincidence at positions $\mathbf{r}_1$ and $\mathbf{r}_2$ at time t in the absence of interferometers. The absolute value squared of this amplitude gives us, up to a prefactor, the probability of this event. The corresponding probability for the time $t - \Delta T$ is the same, in accordance with the fact that the emission time is quantum mechanically uncertain. In the presence of an interferometer, the two amplitudes add up. The first relates to the situation when both photons took the short path through the respective interferometer; the second to the situation when they took the long path (and started correspondingly earlier). The second differs from the first simply by a phase factor determined by the sum of the path differences which the signal and the idler waves accumulated on the long path. The two path differences have a dispersion due to the large bandwidths of the waves. Their sum, however, has a sharp value; it is exactly the path difference the pump wave would accumulate when fictitiously going through both interferometers.

According to Equation (11.19), we find an interesting interference phenomenon: changing one of the phases Φ_s, Φ_i over several periods, the coincidence rate will oscillate sinusoidally. We can observe "interference fringes" analogous to the interference patterns known from conventional interference experiments. Of particular interest is that the fringe visibility attains unity. Of importance also is that the interference pattern in Equation (11.19) is a specifically quantum mechanical effect because it results from the signal and idler photon entanglement. In fact, it was shown by several authors that in a classical description the fringe visibility has the maximum value 50%. Bell type arguments, as sketched in Section 11.3, also apply to the Franson experiment. Assuming a local hidden variables theory, we find, as a Bell inequality, the statement that the fringe visibility cannot exceed the value $1/\sqrt{2} = 70.7\%$. The analyzed experiment thus offers the possibility to test this Bell inequality. Indeed, Kwiat, Steinberg and Chiao (1993) observed a fringe visibility of 80.4%, and thus demonstrated a clear violation of the Bell inequality.

Finally, let us stress once more the interesting point that the linewidths of the individual photons, and therefore also their coherence lengths, do not appear in Equations (11.18) and (11.19) and so do not play a role in the experiment. Instead, the linewidth of the pump wave is physically relevant. It did not appear in our derivation because we assumed a monochromatic pump wave. This is an idealization, which is justified only for times shorter than the coherence time of the pump wave. When ΔT becomes larger than this time, the fringe visibility will decrease and finally vanish. The characteristic coherence length in the Franson experiment is the coherence length of the pump wave, which sounds a bit mysterious because the pump wave survives only in the "memories" of the photons which actually take part in the experiment.

12

Quantum cryptography

12.1 Fundamentals of cryptography

The essence of the Einstein–Podolsky–Rosen experiment analyzed in the preceding chapter is our ability to provide two observers with unpolarized light beams, consisting of sequences of photons, which are coupled in a miraculous way. When both observers choose the same measurement apparatus – a polarizing prism with two detectors in the two output ports, whereby the orientation of the prism is set *arbitrarily* but identically for both observers – their measurement results are *identical*. The measurement result, characterized, say, by "0" and "1," is a genuine random sequence – the quantum mechanical randomness rules unrestricted – from which we can form a sequence of random numbers using the binary number system. The experimental setup thus allows us to deliver simultaneously to the two observers an *identical* series of random numbers. This would be, by itself, not very exciting. Mathematical algorithms can be used to generate random numbers, for example the digit sequence of the number π, which can be calculated up to an arbitrary length. Even though we cannot be completely sure that such a sequence is *absolutely* random, such procedures are sufficient for all practical purposes. The essential point of the Einstein–Podolsky–Rosen experiment is that "eavesdroppers" cannot listen to the communication without being noticed by the observers. When eavesdroppers perform an observation on the photons sent, they inevitably destroy the subtle quantum mechanical correlations, and this damage is irreparable. The Einstein–Podolsky–Rosen correlations can become the basis of an *eavesdropper safe* transmission of an identical sequence of random numbers to two remote observers, traditionally called Alice and Bob. This is exactly what is needed for cryptography, namely encoded, absolutely secret, information transmission. Let us go into details.

The basic principle of cryptography is the coding of plain text by replacing letters by other symbols (or other letters). Achieving this by constantly using the same rule, replacing, for example, the letter "a" always by the same symbol, would make it simple for an unauthorized person to "break the code." (This concept was described in a magnificent and literary way in the short story "The Gold Bug" by Edgar Allen Poe.) After guessing the language in which the text might be written, it is easy to succeed with a frequency analysis: first find the most frequent symbol in the coded text and identify it with the most frequent letter used in the language, then look for the second most frequent symbol, and so on. Much also can be achieved by guesswork.

An absolutely secure protection against decoding is possible using a sequence of random numbers as the key (the sequence should be as long as the text). The key itself must be kept secret; in contrast to the ciphered text and the encoding and decoding procedures, the key must be available only to the authorized users. The coding itself is achieved with the help of the Vernam algorithm: first the letters of the plain text are converted into numbers (usually in binary representation), using strict rules given in the form of a table, for example. Then the random numbers of the key are added (modulo 2) to the converted text one after the other so that each random number is used only once. The result is the ciphered text, the cryptogram. The deciphering is performed simply by reversing the performed transformation: the random numbers of the key are subtracted (modulo 2) from the encoded text, and the result is translated into plain language using the known table. Because a random key was used, the cryptogram is free of any system and the code is therefore completely safe. What is absolutely vital is the secrecy of the key. The key can be given, for example, to a spy who reports the encrypted message to his authority via short wave communication, with the strict instruction to destroy the key immediately after its use. The danger is, however, that the spy will be caught together with the key. A better variant is the use of an eavesdropper safe communication channel for the key transmission. A method for implementation is offered by the subtle peculiarities of quantum mechanics, namely correlations of the Einstein–Podolsky–Rosen type, as discussed in the introduction to this chapter.

From a practical point of view, these things cannot be taken seriously. The theoretician, however, cannot resist the attraction of such considerations which underline the unexpected possibilities hidden in quantum mechanics. It is almost a rule in physics that such "purely academic" considerations lead eventually to practicable solutions. We will return to this later. Let us first discuss in more detail how the use of an Einstein–Podolsky–Rosen channel can protect us against unwanted eavesdropping.

12.2 Eavesdropping and quantum theory

We emphasize that the generation and communication of a secret key in the form of a random number sequence has nothing to do with information transmission.

The key itself does not have any meaning. What strategy can be developed by Alice and Bob to protect their communication against eavesdropping (perhaps we ought to think of the eavesdropper as a lady named Eve)? A possible variant is the following (Bennett, Brassard and Ekert, 1992): Alice and Bob use the same measurement device to measure linear polarization, namely a polarizing prism with a detector in each output. They note a one or a zero, dependent on whether the first or the second detector registered the incident photon. We assume the photon pairs to be emitted at discrete and equidistant times, which can be accomplished by pumping a non-linear crystal with a train of pulses. Alice and Bob therefore know when to expect the arrival of a photon. In addition, they agree to change constantly the settings of their polarization prisms – mutually independently – in a statistically arbitrary way. They will choose between the following two orientations: (a) horizontal–vertical in which the outgoing extraordinary beam is vertically and, accordingly, the ordinary beam horizontally polarized, and (b) rotated by 45° in such a way that the polarization directions of the extraordinary (and therefore also that of the ordinary) beams are the same for both observers.

Thanks to the existence of quantum mechanical correlations between the polarization directions of both photons, this guarantees that the observers, in case they have *accidentally* chosen the same orientation of the polarizing prism, find the same measurement result (interpreted as "0" or "1"). This does not apply for different settings. According to Equation (11.4), when Alice measures a one, for example, Bob's apparatus will indicate in half of the cases a one and in the other half a zero. This is easy to see: imagine, for example, that one of the observers, say Alice, makes a measurement slightly earlier (because she is nearer to the source). The ensemble of photons received by Bob is vertically polarized when it was selected, according to the criterion that Alice registered a vertically polarized photon in each case. When Bob now sends such selected photons onto a 45° degree rotated polarizing prism, this light will be split (from the classical point of view), in equal shares, into two partial beams with polarization directions rotated by 45° or 135° from the vertical. Quantum mechanically, this means that both detectors respond with 50% probability.

What happens when Eve, our eavesdropper, is tapping the line? A strategy that would be a reasonable one for Eve is the following: she uses the same apparatus as Alice and Bob and changes accidently the two orientations (a) and (b). Then either of the two following situations can happen. First, Eve chose by chance the same orientation as Alice, and so finds the same result as Alice. To avoid being detected, Eve sends afterwards an identically polarized photon, as a replacement for the photon consumed by the measurement, to Bob. A cloning of the first photon is not possible! Secondly, Eve chose a polarization different from Alice's, so when she follows the above strategy she distorts the original photon. When the photon was vertically polarized, for example, the replacement will be polarized either 45°

or 135° from the vertical. In both cases, Bob, when he chooses the same orientation as Alice, will find in 50% of the cases a horizontally polarized photon, which is impossible when everything is correct. Such errors allow Alice and Bob to find out that an eavesdropping attack has taken place.

Alice and Bob proceed in detail as follows. After they have completed their measurements, they inform one another about the measurement settings chosen at the respective times. They eliminate as useless all the data obtained for different orientations. To offset the losses in transmission (for example in glass fibers) and possible inefficiencies of the detectors, they communicate publicly the times at which photons should have arrived (and the settings of their apparatuses were identical) but did not, and they discard these data. What remains is a sequence of measurement signals which are, according to quantum mechanics, *identical* for both observers, provided no eavesdropping occurred. To be sure that this was the case, Alice and Bob select, with the help of a random but common key, several of their data (for example the 23rd, 47th, 51st, etc. value in their cleaned sequence) and inform one another about them using public communication channels. Finding the expected exact coincidence, they are satisfied, and they eliminate the now publicly known data from their lists which form the desired absolutely secret key.

The security of the key, as discussed previously, is guaranteed from the beginning by the impossibility of objectifying the polarization properties of individual photons. If Eve could determine through a measurement on a single photon its "objective" polarization, she would be able to send to Bob an exact copy and she would be undetectable. Such things are basically possible in classical theory! Although we made use of the specific quantum mechanical correlations á la Einstein, Podolsky and Rosen, we actually do not need them![1] Instead of starting from correlated photon pairs and leaving "her" photon to choose between two orthogonal polarization directions, Alice can generate the photons she sends to Bob and set their polarization state at will. The secret transmission of the key can thus be realized much more simply (Bennett *et al.*, 1992) in the following way. Alice has a light source at her disposal which generates at given times, say, a vertically polarized photon. Alice changes the polarization state of the photon at will with the help of two Pockels cells, thereby choosing the polarization directions to be either 0°, 45°, 90° or 135° (with respect to the vertical). Bob performs measurements in the same way as before. He communicates to Alice, through a public channel, the times at which

---

[1] These correlations could be used for an additional test. Alice and Bob rotate their polarization prisms not only by 45° but also by 22.5° (also purely random). From the measured data they can, through communication via public channels, calculate coincidence counting rates for a difference between the polarization directions of 22.5° and 67.5° and thus test the Bell inequality. When it is violated, as is predicted by quantum theory (see Section 11.3), they can be sure that no eavesdropper was at work.

he detected a photon and with which setting of his apparatus. Alice compares it with her record of polarization settings, and informs Bob which data to keep.

A prototype of such a transmission has already been achieved (Bennett *et al.*, 1992). The vertically polarized primary photons were replaced, for practical reasons, by weak flashes of light emitted from a diode, and these were linearly polarized by a polarizing filter. When the flashes contain more than one photon, eavesdropping is possible: Eve splits the signal with the help of a beamsplitter, makes a measurement on one of the beams and sends the other unchanged to Bob. This danger can be avoided by severely weakening the primary signal to an average photon number of less than one (for instance one-tenth). This reduces dramatically the transmission rate – in about 90% of the cases no photons will be registered at all – but the probability of detecting two photons in one flash (one by Eve and one by Bob) will be negligibly small.

Working with such extremely weak light pulses where, in most cases, the pulses are empty (in the sense that the detectors do not respond), the experimental setup can be simplified further. Instead of converting the polarization into binary digits (0 and 1), we can work with other photon properties, such as the frequency. For example, Alice sends to Bob red and green light pulses in a random sequence, which contain, on average, much less than one photon. Under such circumstances, Eve will detect very few signals when she chooses to use a beamsplitter for eavesdropping – she will not detect all those which Bob detects. Eve will obtain only a very fragmentary knowledge of the key. Bob informs Alice, in public, about those times when he detected photons but not about their color sequence, which forms the secret key. We have arrived again at photons whose behavior is governed by chance, thus illustrating the break with classical concepts.

Glass fibers are good candidates for signal transmission using polarized photons over long distances; distances of more than 20 km have already been surpassed. A weak point of quantum cryptography is its low transmission rate. Therefore, the encoding of secret messages is, in practice, performed in a different way, namely using so-called "one-way functions" – these are easily calculated (quickly) in one direction used for the encoding, whereas the decoding (the calculation of the reverse function) is extremely time consuming. A representative example is the decomposition of a large number into prime factors. The prime factors represent the secret key and are known only to the authorized user. From the knowledge of the public key obtained by multiplication of the factors, it is not possible to recover the factors in a realistic time, even with the most advanced computers. The procedure is as follows: Bob chooses randomly two prime numbers and multiplies them. He sends the product to Alice, and she uses the result as the key for coding. The decoding is only possible if the factors are known, and they are known only to Bob. Bob is therefore the only receiver able to read the message transmitted from Alice.

13

Quantum teleportation

13.1 Transmission of a polarization state

The word "teleportation" comes from parapsychology and means transportation of persons or things from one place to another using mental power. It was taken over into science fiction literature, where the transport is imagined to take place instantaneously. However this is still to be invented, and is surely nonsense – relativity theory teaches us that the velocity of light is the upper bound for the motion of an object. Nevertheless, teleportation has occupied a firm place in our fantasies, and when renowned quantum physicists (as has happened) use this word, they can be sure to attract attention. So, what is it all about? The basic idea is that it is not necessary to transport material constituents (ultimately the elementary particles). The same particles already exist at other places; we "simply" need to put them together in the right way. To do this, we need a complete set of building instructions, and this is, according to quantum theory, the quantum mechanical wave function representing the maximum information known about an object. We could imagine the wave function measured on the original system, then transmitted via a conventional (classical) information channel to another place and there used for system reconstruction. Unfortunately, the first step, the determination of the wave function on a *single system*, is impossible (see Section 10.3). However, quantum mechanics offers us another "magic trick." It is first of all important to realize that Alice, who wants to teleport, does not need to know the wave function: it is the *transmission* of the wave function that needs to be successful. That this is feasible will be shown in an example. There is, however, a serious drawback to this. The receiver, Bob, cannot simply imprint the received wave function on the matter (or radiation) present at the site; rather, his system must somehow be "intimately related" to the system at Alice's site. In other words, the two systems must be entangled. Careful preparations are necessary. The teleportation relies on the fact that Alice's measurement on the original system also influences the partner system. What is really sent to Bob is the result of such a measurement.

Let us analyze the promised example. Bennett *et al.* (1993) contrived a procedure to teleport the polarization state of a photon. Let us assume the photon (labeled 1) to be in a polarization state described by the superposition

$$|\varphi_1\rangle = \alpha|x\rangle_1 + \beta|y\rangle_1 \tag{13.1}$$

(which Alice need not know). The states $|x\rangle$ and $|y\rangle$ describe, as before, the polarization state of a photon linearly polarized in the x and y direction, respectively, and α, β are arbitrary complex numbers satisfying the normalization condition $|\alpha|^2 + |\beta|^2 = 1$. The state described by Equation (13.1) can describe any type of polarization, in general elliptical.

The "vehicle" used for teleportation is a pair of polarization entangled photons in the state

$$|\Psi_{23}^-\rangle = \frac{1}{\sqrt{2}}(|x\rangle_2|y\rangle_3 - |y\rangle_2|x\rangle_3) \tag{13.2}$$

(see Equation (11.4)). The pair is generated by a source which sends photon 2 to Alice and photon 3 to Bob. (The indices 1, 2 and 3 refer to three different modes of the radiation field.) The goal is to put photon 3 into the state of the original photon 1 as in Equation (13.1). To accomplish this task, Alice and Bob must become active. Alice must perform a measurement on photon 1 including photon 2 (resulting in a destruction of both photons) and she must communicate the result to Bob via a classical communication channel. Bob learns in this way which manipulation he has to carry out on photon 3 to obtain a duplicate of photon 1.

The theoretical background to this procedure consists in a decomposition of the output state of the whole system of three photons in terms of Bell states (introduced in Section 11.1) of photons 1 and 2 (which form a complete basis of the Hilbert space of both photons). A short calculation leads to

$$|\varphi_1\rangle|\Psi_{23}^-\rangle = \tfrac{1}{2}\big\{(\alpha|y\rangle_3 - \beta|x\rangle_3)|\Phi_{12}^+\rangle + (\alpha|y\rangle_3 + \beta|x\rangle_3)|\Phi_{12}^-\rangle$$
$$+ (-\alpha|x\rangle_3 + \beta|y\rangle_3)|\Psi_{12}^+\rangle + (-\alpha|x\rangle_3 - \beta|y\rangle_3)|\Psi_{12}^-\rangle\big\}. \tag{13.3}$$

We see how to proceed. Alice must ensure, through proper measurement, that the system consisting of photons 1 and 2 is transformed into one of the Bell states. The complete wave function given in Equation (13.3) is thus reduced to that part which is in agreement with the measurement result. Photon 3, arriving at Bob's site, is described by the state in the corresponding round brackets of Equation (13.3). Looking more closely at these terms, we find that the fourth is identical to the state of the original photon 1, while the others can be converted into this state by a sign change of α or β, respectively, and/or through the permutation of the states

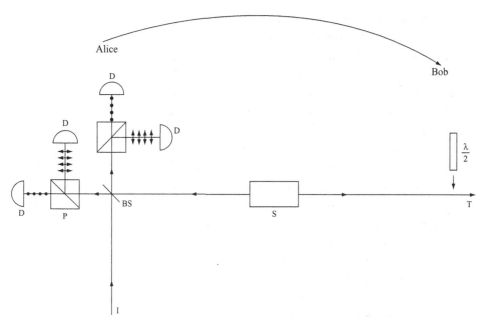

Fig. 13.1. Teleportation of a polarization state. Alice communicates to Bob her measurement results. S = source of entangled photon pairs; BS = beamsplitter; P = polarizing prism; D = detector, I = input signal, T = teleported signal.

x and y. The sign change can be experimentally achieved using a half-wave plate oriented in the y direction, and the interchange of x and y is achieved through a 45° rotated half-wave plate. When Alice communicates which Bell state she measured, Bob knows exactly what to do to prepare an exact copy of the original photon.

The difficult part is on Alice's side: she mixes the original photon, with the help of a semitransparent mirror, with photon 2 to distinguish experimentally between the Bell states (Fig. 13.1).

Polarizing prisms are positioned into the reflected and transmitted beams with detectors in each of the output ports (the orientations of the prisms coincide with the x and y directions). A detailed analysis (Braunstein and Mann, 1995) shows that there are no problems distinguishing between the states $|\Psi_{12}^-\rangle$ and $|\Psi_{12}^+\rangle$: in the first case we detect at each prism one photon and the two photons have different polarizations (in the x and y directions, respectively). In the second case, two differently polarized photons are detected at the first or the second polarizing prism. Unfortunately, it is not possible to distinguish between the two states $|\Phi_{12}^+\rangle$ and $|\Phi_{12}^-\rangle$ with the present apparatus; two photons leave jointly any of the four output ports of the polarizing prisms, with the same probability. The teleportation therefore succeeds only in 50% of all cases. Just to demonstrate the possibility of teleportation, the experimentalist can decide to give away an additional 25% and be content with the detection of the Bell state $|\Psi_{12}^-\rangle$; according to Equation (13.3) this spares Bob the effort of performing any manipulations on his photon. As stated before, the Bell state $|\Psi_{12}^-\rangle$ is the only one which is detected by registering a photon at each of the prisms. This property is thus sufficient for its identification. This has also reduced Alice's experimental effort: the polarizing prisms are not required and the photons can be sent directly onto the detectors.

Experimentally, the quantum teleportation process proceeds as follows (with Alice using the apparatus described in Fig. 13.1). Finding a coincidence, Alice knows that (since the projection is an instantaneous process) at any instant photon 3 is in the same state as the original photon 1. Bob is ignorant of this so Alice has to communicate to him the happy event. This can happen only at the speed of light at maximum, which saves causality. When Alice does not find a coincidence, the teleportation failed. Taking into consideration the usually low detector efficiency, we see that the number of actually measured coincidences drops further, and that the whole procedure, realized for the first time by the Zeilinger group at Innsbruck (Bouwmeester *et al.*, 1997), is inefficient. Notwithstanding this, the experiment confirmed the prediction of the theory that a teleportation actually took place, and thus the concept of teleportation at least was proven.

What is the physical basis behind the teleportation concept? The basic mechanism is projection. Therefore, only one polarization state, from all those possible for photon 3, will become real (factual). An ensemble of photons of type 3 are in a completely unpolarized state (as long as measurements are performed only on photon 3 alone). Using corresponding polarization optics, photons with arbitrary polarization state can be prepared from this ensemble. The measurement on photon 3 can be replaced by a measurement on photon 2 because the two photons are strongly entangled. This was explained in Section 11.3. Because we are interested only in principles, let us for simplicity limit ourselves to linear polarization (of different orientations). The teleportation process simplifies considerably when Alice knows the polarization of photon 1 (this is the case in the experiment anyway); i.e. the apparatus used for preparation is a polarizer rotated by an angle θ. Alice can then proceed in the following way: she lets photon 2 fall on a polarization prism also rotated by the angle θ, with a detector in each output. Depending on which of the detectors responds, she communicates to Bob whether the photon is already in the desired state or its polarization is rotated by $90°$, and Bob can easily undo this.

The important point to be considered in teleportation is that the state of photon 1 is not known, which requires the projection on the Bell states, and this is experimentally more demanding. In any case, the teleportation of a given state is, in principle, the purposeful "crystallizing" of a prescribed state from a kind of potential state, in which the former – as one among many – is latent as a possibility. When quantum teleportation confuses us it means only that we did not really understand the reduction of the wave packet.

Our astonishment increases when we analyze the case of a photon (photon 1) which is not isolated (with a wave function of its own) but is entangled with another photon (photon 0), which is somewhere in the world. From the linearity of the teleportation process, we can immediately conclude that the entanglement will be transferred onto photon 3. The destiny of a particle (photon 0) will be, as soon as it loses its partner (photon 1), tied in a mysterious way to a completely strange particle (photon 3), "until death do they not part."

However surprising the accomplishments of quantum mechanics seem in describing the teleportation process – we see the fantastic potential of quantum mechanical entanglement – a practical person will oppose our description, saying that we are making things unnecessarily complicated. Alice could, for example, send the original photon via a glass fiber directly to Bob. Such a criticism will be void, however, when we replace photons by atoms. When Alice and Bob are separated by a large distance, it would take a long time (compared with the propagation at the velocity of light) before the original atom arrived at Bob's site. Using teleportation, however, we could transfer the quantum mechanical state of the atom much faster (in principle, with the speed of light). We must just take care that the vehicle of teleportation, the entangled atomic pair, started early enough such that one atom arrives at the same time as the original atom at Alice's site and the other atom simultaneously at Bob's site. Indeed, the discussed teleportation scheme can be transferred without any problems to atoms, which, under certain experimental conditions, can be idealized as two-level systems. The two orthogonal polarization states are then replaced by the two energy eigenstates. Experiments with atoms excited to Rydberg states look promising in this respect. Recently, two atoms have been entangled (in analogy to Equation (11.3)) through coherent energy exchange within a non-resonant empty cavity (Osnaghi *et al.*, 2001).

13.2 Transmission of a single-mode wave function

A natural question arises as to whether quantum teleportation is also feasible in the case of continuous variables. The two-dimensional Hilbert space appropriate for polarization is replaced by an infinite-dimensional one. This problem has been solved theoretically (Vaidman, 1994; Braunstein and Kimble, 1998). In addition, the proposed procedure was experimentally realized (Furusawa *et al.*, 1998). The system considered is one mode of the radiation field (in the following denoted again by the index 1). It is its, in principle unknown, quantum state which should be transferred by Alice onto the wave arriving at Bob's site. A suitable vehicle was found to be an entangled two-mode state of the type used already by Einstein, Podolsky and Rosen (1935) in their sophisticated criticism of quantum mechanics. It is a state of two particles with highly correlated position and momentum variables. The total momentum as well as the position difference have sharp values. Identifying the position and momentum with the quadrature components of the field (see Section 9.1), we can easily take the step from a mechanical system to a two-mode system.

In fact, there exist known methods of non-linear optics that allow us to prepare approximately quantum states of two modes 2 and 3 with quadrature components satisfying the relations

$$x_2 = -x_3, \quad p_2 = p_3. \tag{13.4}$$

The quadrature components themselves fluctuate, but the fluctuations of one wave find their counterpart, according to Equation (13.4), in the other wave. This radiation is generated by mixing two beams (with the help of a 50:50 beamsplitter), one of which is strongly squeezed in the x component and the other in the p component (see Section 9.2). (The delta function-like correlations (Equation (13.4)) are obtained only in the limit of infinitely large squeezing.)

The teleportation of wave 1 takes place in the following manner. Alice mixes wave 1 with wave 2 of the vehicle using a beamsplitter and measures the x component on one and the p component on the other output (Fig. 13.2) using the homodyne technique (see Section 9.3).

Let x^0 and p^0 be the measured values. The quantum mechanical description (Braunstein and Kimble, 1998) shows that the state of Bob's wave is readily described in the Wigner formalism: its Wigner function $W_{out}(x_3, p_3)$ is related to the Wigner function $W_{in}(x_1, p_1)$ of the original wave 1 through a simple displacement in phase space (x, p plane) by $-\sqrt{2}x^0$ in the x direction and by $\sqrt{2}p^0$ in the p direction. Thus the following relation holds:

$$W_{out}(x_3, p_3) = W_{in}(x_3 + \sqrt{2}x^0, p_3 - \sqrt{2}p^0). \tag{13.5}$$

To make the result plausible, let us forget for a moment quantum theory and look at the problem from a classical point of view. The Wigner function is then considered as a classical distribution function. We can think of the quadrature components of the three waves at each time – more precisely speaking in each time interval whose length is determined by the coherence time – as having a defined (not predictable) value x_1, p_1, etc. The beamsplitter used by Alice causes the quadrature components of the outgoing waves x', p' and x'', p'', and those of the

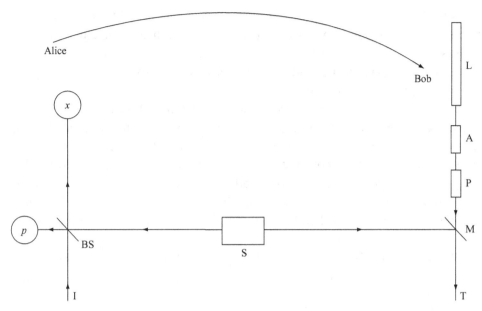

Fig. 13.2. Teleportation of a single-mode wave function. Alice measures the quadrature components x and p and communicates the results to Bob. Using the received values, Bob changes the amplitude and the phase of a laser beam. S = source of an entangled two-mode state; BS = beamsplitter; M = strongly reflecting mirror; L = laser; A = amplitude modulator; P = phase modulator; I = input signal; T = teleported signal.

incident waves x_1, p_1 and x_2, p_2 to be related as follows:

$$x' = \frac{1}{\sqrt{2}}(x_1 + x_2), \qquad p' = \frac{1}{\sqrt{2}}(p_1 + p_2);$$

$$x'' = \frac{1}{\sqrt{2}}(-x_1 + x_2), \qquad p'' = \frac{1}{\sqrt{2}}(-p_1 + p_2). \tag{13.6}$$

(These relations are simply Equation (15.48) for the special case $r = t = 1/2$ rewritten into real and imaginary parts.) When Alice measures x' with the result x^0 and p'' with the result p^0, we have

$$x^0 = \frac{1}{\sqrt{2}}(x_1 + x_2), \qquad p^0 = \frac{1}{\sqrt{2}}(-p_1 + p_2). \tag{13.7}$$

Taking into account also the correlations in Equations (13.4) for waves 2 and 3 we can proceed as follows:

$$x_1 = x_3 + \sqrt{2}x^0, \qquad p_1 = p_3 - \sqrt{2}p^0. \tag{13.8}$$

This implies strong correlations between the original wave and Bob's wave: when the first wave (randomly) takes the values x_1 and p_1, then in the latter wave the values x_3 and p_3 satisfying Equations (13.8) will be realized. The distribution of the quadrature components of the original wave is thus transferred to Bob's wave in the form of Equation (13.5).

Let us return to the *quantum mechanical* prediction in Equation (13.5). In order to reproduce the state of wave 1, the displacement of the Wigner function must be undone. Fortunately, there exists a procedure for this: we have to mix wave 3 with a wave in a coherent state (Glauber state). The mirror used for this task has to be highly reflective (large coefficient of reflectivity r) and correspondingly weakly transparent (small transmitivity coefficient t) so that it will reflect almost all of wave 3 (and hardly change it). The coherent wave becomes superposed with the reflected wave after having passed through the mirror, whereby its Wigner function becomes displaced. In passage, the amplitude of the coherent wave is weakened by a factor of $\sqrt{t}$. We have to compensate for this by increasing the input amplitude by the factor $1/\sqrt{t}$. Taking into consideration the relation $\alpha = (1/\sqrt{2})(x + \mathrm{i}p)$ between the quadrature components and the coherent state complex amplitude α (Section 15.2), Bob has to use for the mixing a Glauber state $|\alpha = (x^0 - \mathrm{i}p^0)/\sqrt{t}\rangle$. (Because of the large amplitude, it is *de facto* a classical wave.) To prepare the wave, Bob needs to know the measurement data found by Alice, which must be communicated to Bob with the speed of light at best. Using optical mixing, Bob undoes the displacement of the Wigner function, and the teleportation was successful!

The two different squeezed light beams in the performed experiment (Furusawa *et al.*, 1998) were generated using a sub-threshold operated optical parametric oscillator with ring type resonator geometry. The oscillator was pumped with two counter-propagating waves coming from the same laser. The input signal (wave 1) originated from the same primary laser; i.e. it was in a Glauber state. The experiment was performed in the continuous regime. The homodyne measurement was thus continuous, and the measured data in the form of difference photocurrents (Section 9.3) were sent directly to Bob's observation station via electric conductors. There they drove, after amplification, amplitude and phase modulators. The modulators prepared the required auxiliary wave from a wave split from the primary laser beam. To validate the teleportation, Furusawa *et al.* introduced a third (fictitious) observer named Victor (verifier) who generated the input signal for Alice and compared it with the teleported signal. The fidelity of the teleportation was found to be $58 \pm 2\%$. The relatively low value is caused, on the one hand, by the insufficient squeezing of waves 2 and 3 and, on the other hand, by propagation losses of the two waves as well as detector inefficiencies. The advantage of this experiment is – contrary to the teleportation of the polarization state of a photon – that the quantum state of each input signal was teleported.

14

Summarizing what we know about the photon

How can we construct a picture of the photon from the wealth of observation material available to us? The photon appears to have a split personality: it is neither a wave nor a particle but something else which, depending on the experimental situation, exhibits a wave- or a particle-like behavior. In other words, in the photon (as in material particles such as the electron) the particle–wave dualism becomes manifest. Whereas classically the wave and the particle pictures are separate, quantum mechanics accomplishes a formal synthesis through a unified mathematical treatment.

Let us look first at the wave aspect familiar from classical electrodynamics, which seems to be the most natural description. It makes all the different interference phenomena understandable, such as the "interference of the photon with itself" on the one hand and the appearance of spatial and temporal intensity correlations in a thermal radiation field on the other (which are obviously brought about by superposition of elementary waves emitted independently from different atoms). It might come as a surprise (at least for those having quantum mechanical preconceptions) that the classical theory is valid down to arbitrarily small intensities: the visibility of the interference pattern does not deteriorate even for very small intensities – the zero point fluctuations of the electromagnetic field advocated by quantum mechanics do not have a disturbing effect – and is valid not only for conventional interference experiments but also for interference between independently generated light beams (in the form of laser light).

The fact that the natural linewidth of the radiation and the mean lifetime of an excited atomic level are related in accordance with the classical theory supports the assumption that the photon is emitted in a continuous process as a wave. A direct consequence of the unlimited validity of the classical description of the interference processes is the perfect functioning of conventional spectrometers, even at very low intensities (we disregard disturbances whose effects are more pronounced at low intensities than at high ones). The spectral decomposition of a single photon

215

is possible (at least in principle). A wave corresponding to a single photon is split into a reflected and a transmitted partial wave in a high resolution Fabry–Perot interferometer. The frequency domains of the partial waves are mutually complementary; i.e. they sum to the spectrum of the original wave. The wave train thus becomes longer. Similarly, a photon in the form of a spherical wave passes partly through a hole in a mirror, while the other part is reflected.

The frequency spectrum of a photon can not only be narrowed but also widened. The latter can be achieved using a fast shutter which can cut out or cut away a part of the wave. According to the Fourier theorem, the spectrum of the resulting wave must be broader than that of the original wave (which is assumed to be coherent). A fundamental problem arises essentially in the process of measurement. It is related to the fact that the detection is based almost entirely on the photoeffect. As a consequence, the photon is lost in the measurement process. The fact that in the case of beamsplitting (of exactly one incident photon) we can detect something at all is already in clear contradiction to classical wave concepts. The decomposition into partial waves should be accompanied by a corresponding splitting of the energy, and hence the detectors should not respond because they require the full energy $h\nu$ of the photon and such an energy is seemingly no longer available. Fortunately, such an energy "dissipation" does not take place. Rather, the photon (as an energetic whole) can be found in one partial beam – for instance in the partial wave running through a Fabry–Perot interferometer.

Thus, the single measurement delivers only very limited information. For example, we find the photon in one of the many output channels of a spectrometer which indicates a certain frequency value to be measured. We cannot measure a frequency spectrum on a single photon; rather, frequent repetition of the experiment enables us to record a spectrum.

The particle character of light manifests itself in the same spectacular way as the wave character. It shows up in the process of spontaneous emission: positioning a detector close enough to the excited atom, the detector clicks in some cases at times that are shorter than the lifetime of the upper level of the atom. According to this the whole energy of the photon $h\nu$ is sometimes already available at the moment when the emission, in the sense of wave generation, has just started! Conversely, absorption acts take place in such a short time that it is not possible for the atom to accumulate the necessary amount of energy from the field if it were distributed there continuously. The energy must be present in an agglomerated form.

The photons conceived as localized energy packets, however, do not possess individuality: it is not possible to trace back the "course of life" of a photon (for example, in a thermal light field) to its moment of birth, even in a Gedanken experiment. If this were the case, there would be no spatial or temporal correlations between events detected by two different detectors. As mentioned previously, propagation processes can be described only with the aid of a wave theory. The wave character of light manifests itself in experiments through a relation between amplitude and probability. The momentary intensity of the light determines the probability of a detector response at the respective position.

Another fundamental property of the photon is its energetic indivisibility, illustrated and discussed in detail using the example of a beamsplitter in Section 7.1. The consequence is that electromagnetic energy cannot be arbitrarily "diluted" – either we find a photon or we do not – what becomes smaller is the probability of the first event when the cross section of the light beam becomes larger through propagation.

It is a characteristic of the particle picture that when it is applied chance also comes into play. This is equally valid for the "instant" of emission as for that of absorption. In addition, in the case of absorption by an ensemble of atoms illuminated by not too strong a wave, only a fraction of the atoms will be excited. The question concerning which atom will "receive" a photon will be decided by "playing dice." Similarly, in beamsplitting – assuming a single incident photon – it is left to chance which detector will receive the photon.

From the randomness of the events, we can see regularities acting which are completely alien to classical mechanics and electrodynamics. In fact, specific quantum mechanical features of natural phenomena are thus revealed. The reason why we have difficulty understanding them, especially in the case of light, is that the "crazy things" that we became used to in the microworld appear now on a *macroscopic* scale. We do not find it exciting that a bound electron in a hydrogen atom should be "smeared out" over a typical scale of 10^{-10} m, but we are reluctant to believe that in the case of the "interference of a photon with itself" the photon is present in both partial beams whereby the spatial separation of the two "halves" approaches meters or kilometers (in principle, there is no limit!). Obviously, we have to accept these concepts, and the recently realized Gedanken experiments of Einstein, Podolsky and Rosen described in Section 11.4 clearly demonstrate that specific quantum mechanical correlations can extend over macroscopic distances. The world is indeed more complicated than one is led to believe, and photons contribute in great part to this.

15

Appendix. Mathematical description

15.1 Quantization of a single-mode field

A single-mode field is characterized by a sharp (circular) frequency ω and a spatial distribution of the electric field strength (given at the initial time $t = 0$; see also Section 4.2). When we assume that the field suffers no interaction and hence is propagating undisturbed, the positive and negative frequency parts of the field amplitude can be written, according to classical electrodynamics, in the form

$$E^{(+)}(\mathbf{r}, t) = F(\mathbf{r})e^{-i\omega t} A, \qquad E^{(-)}(\mathbf{r}, t) = F^*(\mathbf{r})e^{i\omega t} A^*. \qquad (15.1)$$

$F(\mathbf{r})$ characterizes the given spatial field distribution (it is a normalized solution of the time independent wave equation) and A is a complex amplitude.

The quantization of the field is formally realized by replacing the classical amplitude A by a (non-Hermitian) operator $\hat{a}$ and accordingly the complex conjugate amplitude A^* by the Hermitian conjugate operator $\hat{a}^\dagger$. Equations (15.1) are thus replaced by

$$\hat{E}^{(+)}(\mathbf{r}, t) = \mathcal{E}(\mathbf{r})e^{-i\omega t}\hat{a}, \qquad \hat{E}^{(-)}(\mathbf{r}, t) = \mathcal{E}^*(\mathbf{r})e^{i\omega t}\hat{a}^\dagger, \qquad (15.2)$$

where the functions $F(\mathbf{r})$ and $\mathcal{E}(\mathbf{r})$ differ only in their normalization.

The operators $\hat{a}, \hat{a}^\dagger$ satisfy the fundamental commutation relation

$$[\hat{a}, \hat{a}^\dagger] = \mathbf{1}, \qquad (15.3)$$

and the energy of the radiation field, or more precisely the Hamiltonian, reads

$$\hat{H} = \hbar\omega\hat{a}^\dagger\hat{a}. \qquad (15.4)$$

From the commutation relation it follows that the operator $\hat{N} \equiv \hat{a}^\dagger\hat{a}$ has the eigenvalues $n = 0, 1, 2, \ldots$, and due to this and Equation (15.4) it deserves the name "photon number operator." Indeed (for a single-mode field) the response probability of a photodetector is proportional to the expectation value of $\hat{N}$. However, for the coincidence measurement it is not simply the expectation value of $\hat{N}^2$; we have to cast the latter operator into a normally ordered form. This means that all the operators $\hat{a}$ have to be placed on the right of the operators $\hat{a}^\dagger$. The probability that two detectors simultaneously respond is proportional to $\langle \hat{a}^{\dagger 2} \hat{a}^2 \rangle$, which, due to Equation (15.3), can be rewritten as $\langle \hat{N}(\hat{N} - 1) \rangle$.

The eigenvectors $|n\rangle$ of the photon number operator form a complete orthogonal system of functions. They describe field states with sharp photon numbers n and are called Fock states. The action of the operators $\hat{a}$ and $\hat{a}^\dagger$ reads as

$$\hat{a}\,|n\rangle = \sqrt{n}\,|n-1\rangle, \quad \hat{a}^\dagger\,|n\rangle = \sqrt{n+1}\,|n+1\rangle \quad (n = 0, 1, 2, \ldots). \tag{15.5}$$

These relations explain why the operators $\hat{a}$ and $\hat{a}^\dagger$ are called photon annihilation and creation operators. Applying the creation operator n times to the vacuum state, we obtain the state with n photons

$$|n\rangle = (n!)^{-\frac{1}{2}}\,(\hat{a}^\dagger)^n\,|0\rangle. \tag{15.6}$$

Starting from Equations (15.2) and (15.6), we conclude that the expectation value of the electric field strength equals zero for states with well defined photon numbers. In addition, from the definition of the photon number operator and the representation in Equation (15.2) of the field strength operator, it follows that the two operators do not commute. As a consequence, there are no states for which the electric field strength and the energy (photon number) are simultaneously sharp.

The description of the electric field with the aid of the creation and annihilation operators is tailored to the particle aspect of the field. The description of processes involving energy exchange such as emission or absorption takes an intuitive and transparent form when the language of creation and annihilation operators is used, as in these processes photons are really created or annihilated.

When the particle aspect of the field is suppressed in favor of the wave aspect, as in the case of light field squeezing or the homodyne detection technique, description through the so-called quadrature components becomes favorable. Let us consider the simple case of a linearly polarized plane wave. According to Equation (15.2), the complex representation of the electric field strength $\hat{E}(\mathbf{r}, t) = \hat{E}^{(-)}(\mathbf{r}, t) + \hat{E}^{(+)}(\mathbf{r}, t)$ takes the form

$$\hat{E}(\mathbf{r}, t) = E_0 \left\{ e^{i(\mathbf{kr} - \omega t)} \hat{a} + e^{-i(\mathbf{kr} - \omega t)} \hat{a}^\dagger \right\}, \tag{15.7}$$

where E_0 is a constant and $\mathbf{k}$ is the wave vector. Using the decomposition of the exponential into real and imaginary parts, we obtain the real valued representation

$$\hat{E}(\mathbf{r}, t) = \sqrt{2} E_0 \left\{ \hat{x} \cos(\omega t - \mathbf{kr}) + \hat{p} \sin(\omega t - \mathbf{kr}) \right\}. \tag{15.8}$$

The two Hermitian operators

$$\hat{x} = (1/\sqrt{2})(\hat{a}^\dagger + \hat{a}), \quad \hat{p} = (1/\sqrt{2})\mathrm{i}(\hat{a}^\dagger - \hat{a}), \qquad (15.9)$$

are the counterparts of the real and imaginary parts of the classical complex amplitude A, and they represent the two quadratures of the field. They are the exact analogs of position and momentum when we compare the single-mode field with a harmonic oscillator. The chosen notation should remind us of this analogy. Using Equation (15.3) it is easy to check the validity of the commutation relation

$$[\hat{x}, \hat{p}] = \mathrm{i}\mathbf{1}. \qquad (15.10)$$

Equation (15.8) has a simple interpretation. The electric field is split into two parts: one component ($\hat{x}$) is in phase with a classical reference wave oscillating as $\cos(\omega t - \mathbf{kr})$ and the other component ($\hat{p}$) is out of phase.

The chosen reference wave was a very special one. Any wave that differs from the considered one by a phase shift of Θ, i.e. whose electric field strength is given by $E_1 \cos(\omega t - \mathbf{kr} - \Theta)$, is equally well suited for our purpose. This is not only of theoretical but also of practical interest for field detection using the homodyne technique (see Sections 9.3 and 15.6). Equations (15.8) and (15.9) can be replaced by

$$\hat{E}(\mathbf{r}, t) = \sqrt{2}E_0 \left\{ \hat{x}_\Theta \cos(\omega t - \mathbf{kr} - \Theta) + \hat{p}_\Theta \sin(\omega t - \mathbf{kr} - \Theta) \right\}, \qquad (15.11)$$

where

$$\hat{x}_\Theta = (1/\sqrt{2})(\mathrm{e}^{\mathrm{i}\Theta}\hat{a}^\dagger + \mathrm{e}^{-\mathrm{i}\Theta}\hat{a}), \quad \hat{p}_\Theta = (1/\sqrt{2})\mathrm{i}(\mathrm{e}^{\mathrm{i}\Theta}\hat{a}^\dagger - \mathrm{e}^{-\mathrm{i}\Theta}\hat{a}). \qquad (15.12)$$

The new quadrature components are related to the old ones through a simple rotation by the angle Θ:

$$\begin{pmatrix} \hat{x}_\Theta \\ \hat{p}_\Theta \end{pmatrix} = \begin{pmatrix} \cos\Theta & \sin\Theta \\ -\sin\Theta & \cos\Theta \end{pmatrix} \begin{pmatrix} \hat{x} \\ \hat{p} \end{pmatrix}. \qquad (15.13)$$

Because this is a unitary transformation, the operators $\hat{x}_\Theta$ and $\hat{p}_\Theta$ also satisfy the canonical commutation relation, Equation (15.10).

From Equation (15.13) it follows that p_Θ is identical to $\hat{x}_{\Theta+\frac{\pi}{2}}$. By changing the angle Θ from 0 to π, we obtain all possible quadrature components. (Including the interval $\pi, \ldots, 2\pi$ will change only the sign of the quadratures.) The quadratures correspond to different measurements, hence we are dealing with a continuous manifold of observables. The amazing fact is that these observables are measurable in a rather simple way (see Sections 9.3 and 15.6).

15.2 Definition and properties of coherent states

Let us find quantum mechanical states of a single-mode field which come as close as possible to the classical states with well defined phases and amplitudes. To this end we require that the electric field strength has minimum fluctuations. In addition we require that the mean energy expressed through the mean photon number has a given value. Without the additional condition we would obtain eigenfunctions of the field strength operators which do not have a finite mean energy.

Let us straight away relax the requirement for the electric field strength. We are content when the electric field strength fluctuations become as small as possible in the time average (over a light oscillation period). We thus wish to minimize

$$\overline{\Delta E^2(\mathbf{r}, t)}^t \equiv \overline{\langle \hat{E}^2(\mathbf{r}, t) \rangle}^t - \overline{\langle \hat{E}(\mathbf{r}, t) \rangle^2}^t = \text{minimum}, \tag{15.14}$$

with the constraint

$$\langle \hat{N} \rangle \equiv \langle \hat{a}^\dagger \hat{a} \rangle = \overline{N} (\text{fixed}). \tag{15.15}$$

Using Equations (15.2) and (15.3), we rewrite Equation (15.14) as follows:

$$\overline{\Delta E^2(\mathbf{r}, t)}^t = 2|\mathcal{E}(\mathbf{r})|^2 \left(\langle \hat{a}^\dagger \hat{a} \rangle - \langle \hat{a}^\dagger \rangle \langle \hat{a} \rangle + \tfrac{1}{2} \right) = \text{minimum}. \tag{15.16}$$

Combining Equations (15.16) and (15.15), we see that minimizing the mean field strength fluctuations is equivalent to maximizing the product $\langle \hat{a}^\dagger \rangle \langle \hat{a} \rangle = \langle \hat{a} \rangle^* \langle \hat{a} \rangle$ (the squared absolute value of $\langle \hat{a} \rangle$) for a fixed mean photon number $\langle \hat{a}^\dagger \hat{a} \rangle$.

For the unknown state we make the ansatz

$$|\psi\rangle = \sum_{n=0}^{\infty} c_n |n\rangle, \tag{15.17}$$

and express the expectation value of $\hat{a}$ through the unknown expansion coefficients c_n. With the help of Equation (15.5) we obtain

$$\langle \psi | \hat{a} | \psi \rangle = \sum_n c_n^* c_{n+1} \sqrt{n+1}. \tag{15.18}$$

The right hand side of Equation (15.18) can be understood as the scalar product of two vectors with components $a_n = c_n$ and $b_n = c_{n+1} \sqrt{n+1}$. The scalar product satisfies Schwarz's inequality

$$\left| \sum_n a_n^* b_n \right|^2 \leq \sum_n |a_n|^2 \sum_n |b_n|^2, \tag{15.19}$$

stating simply that the modulus of the scalar product cannot exceed the product of the lengths of the two vectors, which is geometrically evident. Applied to our case we obtain the inequality

$$|\langle\psi|\hat{a}|\psi\rangle|^2 \le \sum_n |c_n|^2 \sum_n |c_{n+1}|^2(n+1) = \overline{N}. \tag{15.20}$$

Fortunately, on the right hand side of this equation the mean photon number appears, which is assumed to be given. The question of when the left hand side of the inequality becomes maximum is easily answered: it occurs only when the two vectors are parallel to one another; i.e. when the equation $b_n = \alpha a_n$ holds, or, in our case,

$$c_{n+1}\sqrt{n+1} = \alpha c_n, \tag{15.21}$$

where α is an arbitrary complex constant. Equation (15.21) is a simple recursion formula, and can be easily solved to obtain the explicit representation of the coefficients c_n, namely

$$c_n = \frac{c_0}{\sqrt{n!}}\alpha^n. \tag{15.22}$$

Satisfying the normalization condition $\sum_n |c_n|^2 = 1$, we find the desired wave function to be

$$|\alpha\rangle = e^{-|\alpha|^2/2} \sum_{n=0}^{\infty} \frac{\alpha^n}{\sqrt{n!}}|n\rangle. \tag{15.23}$$

This relation completely defines the coherent states. They are very often named after R. Glauber, who was the first to realize their usefulness for the quantum mechanical description of the radiation field.

The definition in Equation (15.23) can be used to derive several interesting formal properties of Glauber states. Applying the photon annihilation operator $\hat{a}$ on the state $|\alpha\rangle$, we find the simple relation

$$\hat{a}|\alpha\rangle = \alpha|\alpha\rangle, \tag{15.24}$$

which is a kind of eigenvalue equation. However, we have to note that $\hat{a}$ is not a Hermitian operator. For the Hermitian conjugate operator $\hat{a}^\dagger$, we obtain another relation, namely

$$\langle\alpha|\hat{a}^\dagger = \alpha^*\langle\alpha|. \tag{15.25}$$

To be precise, we should say that the Glauber states are right hand side eigenvectors of $\hat{a}$ and left hand side eigenvectors of $\hat{a}^\dagger$. However, they are not eigenvectors of any Hermitian operator and therefore we cannot ascribe to them any precise values of any observable, and, as a consequence, such states cannot be produced by a measurement process, even in a Gedanken experiment.

Equations (15.24) and (15.25) allow a very elegant calculation of quantum mechanical expectation values. Above all, this is true for the normally ordered operators that we are dealing with when we describe measurements performed with photodetectors (see Sections 5.2 and 15.1). From Equations (15.24) and (15.25), we derive first the simple relations

$$\langle\alpha|\hat{N}|\alpha\rangle = \langle\alpha|\hat{a}^\dagger\hat{a}|\alpha\rangle = |\alpha|^2, \tag{15.26}$$

$$\langle\alpha|\hat{a}|\alpha\rangle = \alpha, \quad \langle\alpha|\hat{a}^\dagger|\alpha\rangle = \alpha^*, \tag{15.27}$$

$$\langle\alpha|\hat{a}^2|\alpha\rangle = \alpha^2, \quad \langle\alpha|\hat{a}^{\dagger 2}|\alpha\rangle = \alpha^{*2}. \tag{15.28}$$

Let us recall that the operator $\hat{a}$ corresponds to the classical complex amplitude A. Because of this, the complex parameter α has the meaning of such an amplitude, and, according to Equation (15.26), the normalization is chosen in such a way that $|\alpha|^2$ indicates the mean photon number. In particular, using Equations (15.2) and (15.26) to (15.28) we obtain

$$\langle\alpha|\hat{E}(\mathbf{r}, t)|\alpha\rangle = \mathcal{E}(\mathbf{r})e^{-i\omega t}\alpha + \mathcal{E}^*(\mathbf{r})e^{i\omega t}\alpha^*, \tag{15.29}$$

$$\langle\alpha|\hat{E}^2(\mathbf{r}, t)|\alpha\rangle = \mathcal{E}^2(\mathbf{r})e^{-2i\omega t}\alpha^2 + \mathcal{E}^{*2}(\mathbf{r})e^{2i\omega t}\alpha^{*2} + 2|\mathcal{E}(\mathbf{r})|^2\left(|\alpha|^2 + 1/2\right), \tag{15.30}$$

where the commutation relation in Equation (15.3) was used in Equation (15.30). Apart from the additional term $|\mathcal{E}(\mathbf{r})|^2$ in Equation (15.30), which represents the contribution from the vacuum fluctuations of the field and hence is not understandable from a classical point of view, the right hand sides expressions of Equations (15.29) and (15.30) are exactly the classical expressions. This implies in particular that the phase of the complex number α corresponds to the phase of the field. However, from the quantum mechanical point of view, it is not a well defined value, i.e. an eigenvalue (of a Hermitian phase operator), but rather it is the center of a phase distribution of finite width.

Let us calculate from Equations (15.30) and (15.29) the dispersion of the electric field strength. We find the same result as that obtained by time averaging, namely

$$\Delta E^2(\mathbf{r}, \mathbf{t}) \equiv \langle\alpha|\hat{E}^2(\mathbf{r}, t)|\alpha\rangle - \langle\alpha|\hat{E}(\mathbf{r}, t)|\alpha\rangle^2 = |\mathcal{E}(\mathbf{r})|^2. \tag{15.31}$$

As the previous discussion clearly demonstrated, the right hand side of Equation (15.31) is really the absolute minimum which the variance of the electric field strength can reach under the constraint of a finite field energy. As already mentioned, $|\mathcal{E}(\mathbf{r})|^2$ represents the contribution from the vacuum fluctuations. It is easy to check that Equation (15.31) holds true also for the vacuum state $|0\rangle$ (this comes as no surprise because the vacuum state can be understood as a limiting case of the Glauber state $|\alpha\rangle$ for $\alpha \to 0$). This leads us to the conclusion that the Glauber states come as close as possible to classical states with well defined amplitudes and phases. The vacuum fluctuations impose an unbeatable limit. They are in the end responsible for the fact that the phase and the amplitude or the energy, respectively, of a light field fluctuate. From Equation (15.23) we learn that the probability of detecting (in the case of a perfect measurement) exactly n photons is given by the Poissonian distribution

$$w_n \equiv |c_n|^2 = \frac{e^{-|\alpha|^2}|\alpha|^{2n}}{n!} = \frac{e^{-\overline{N}}\overline{N}^n}{n!}, \tag{15.32}$$

with $\overline{N} = |\alpha|^2$ being the mean photon number. The photon number variance then follows as

$$\Delta N^2 \equiv \langle \hat{N}^2 \rangle - \langle \hat{N} \rangle^2 = \overline{N}. \tag{15.33}$$

In the limit $\overline{N} \to \infty$, the *relative* variance of the photon number goes to zero as $1/\overline{N}$, and we come arbitrarily close to a classical state with sharp energy, as expected from the correspondence principle.

The coherent states have a drawback, however, they are not mutually orthogonal. Their scalar product is

$$\langle \alpha | \beta \rangle = e^{\alpha^* \beta - (|\alpha|^2 + |\beta|^2)/2}, \tag{15.34}$$

which gives for the squared absolute value

$$|\langle \alpha | \beta \rangle|^2 = e^{-|\alpha - \beta|^2}. \tag{15.35}$$

Due to this property they cannot be eigenvectors of any Hermitian operator. However, they form a complete (more precisely, overcomplete) system; i.e. the unit operator can be decomposed in the form

$$\mathbf{1} = \frac{1}{\pi} \int d^2\alpha \, |\alpha\rangle \langle \alpha|, \tag{15.36}$$

where the integration is over the whole complex α plane.

There exists a large class of quantum mechanical states that are distinguished by the fact that the corresponding density operator can be represented in the special form

$$\hat{\rho} = \int d^2\alpha \, P(\alpha) |\alpha\rangle \langle \alpha|, \tag{15.37}$$

where $P(\alpha)$ is a real function, not necessarily positive. When we are able to satisfy Equation (15.37) with a non-pathological function $P(\alpha)$, we say there exists a Glauber P representation of the density matrix $\hat{\rho}$.

Equation (15.37) allows us to work out a correspondence between the classical and the quantum mechanical description. We calculate the mean values of normally ordered products of creation and annihilation operators using Equation (15.37). Using Equations (15.24) and (15.25), we can replace $\hat{a}$ by α and $\hat{a}^\dagger$ by α^*, and the following simple relations hold:

$$
\begin{aligned}
\langle \hat{a}^{\dagger k} \hat{a}^l \rangle &\equiv \mathrm{Tr}(\hat{a}^{\dagger k} \hat{a}^l \hat{\rho}) \\
&= \int \mathrm{d}^2\alpha \, P(\alpha) \langle \alpha | \hat{a}^{\dagger k} \hat{a}^l | \alpha \rangle \\
&= \int \mathrm{d}^2\alpha \, P(\alpha) \alpha^{*k} \alpha^l \quad (k, l = 0, 1, 2, \dots).
\end{aligned}
\tag{15.38}
$$

The last expression is just the classical ensemble average when we interpret $P(\alpha)$ as a classical distribution function. However, in this case we assume beforehand that $P(\alpha)$ is *positive definite* because a negative probability makes no sense.

We arrive at the following general conclusion. The quantum mechanical and the classical description lead to *identical* predictions when, first, we restrict ourselves to normally ordered operators and, secondly, the density operator has a positive definite P representation. For states that do not satisfy these conditions – the Fock states are a simple example – we can expect non-classical behavior.

15.3 The Weisskopf–Wigner solution for spontaneous emission

In spontaneous emission, an excited atom emits light at different frequencies and into all possible directions. From a quantum mechanical point of view this means that the atom is coupled virtually to all possible modes of the radiation field. We choose the light modes as linearly polarized running plane waves, characterized by an index μ that stands for the polarization direction and the wave vector. In addition, we idealize the atom as a two-level system with an upper level b and a lower level a. We assume the atom to be pinned to the origin of the coordinate system. This means that we assume the mass of the atom to be sufficiently large so that a good spatial localization of the atom associated only with negligibly small fluctuations of its velocity around zero is guaranteed – due to Heisenberg's uncertainty relation a small spatial uncertainty is always connected with a considerable momentum uncertainty. The recoil of the atom from the emission can also be neglected. The wave function of the system formed by the atom and the radiation field can then be written in the form

$$
|\psi(t)\rangle = f(t)|b\rangle|0, 0, \dots\rangle + \sum_\mu g_\mu(t)|a\rangle|0, 0, \dots, 0, 1_\mu, 0, \dots\rangle,
\tag{15.39}
$$

with $|0, 0, \dots\rangle$ being the vacuum state of the field and $|0, 0, \dots, 0, 1_\mu, 0, \dots\rangle$ denoting a field state with just one mode μ excited with exactly one photon.

The motivation for Equation (15.39) is the following. We start from the idealized situation that the atom is excited with certainty and the field is completely "empty". This is a pure state, characterized by the initial values

$$f(0) = 1, \quad g_\mu(0) = 0, \tag{15.40}$$

and it is known that the whole system will remain for all time in a pure state, which justifies the ansatz of the wave function. In Equation (15.39) the energy conservation law was taken into account. The atom emits a photon when it goes from b to a and absorbs a photon in the reverse process. Even though this sounds plausible, it is still an approximation, the so-called rotating wave approximation. Actually, the exact interaction operator (in the dipole approximation) also contains terms that contradict the energy conservation law (the transition of the atom from b to a is associated with the absorption of a photon, etc.). The additional terms do not play a significant role in spontaneous emission; however, they have a serious physical meaning.

Writing down the Schrödinger equation for the whole system (in the dipole and rotating wave approximations), we obtain a coupled linear set of equations for the unknown functions $f(t)$ and $g_\mu(t)$. This set can be solved exactly using the Laplace transformation method; however, the back transformation causes difficulties. It is not possible to find a closed form solution, but the following formula, known as the Weisskopf–Wigner solution (Weisskopf and Wigner, 1930a, b), represents a good approximation

$$f(t) = e^{-\Gamma t/2}, \tag{15.41}$$

$$g_\mu(t) = \frac{\langle a, 0, 0, \ldots, 1_\mu, 0, \ldots \, | \hat{H}^{\mathrm{I}} | b, 0, \ldots \rangle}{\hbar} \frac{e^{-\mathrm{i}(\omega_\mu - \omega_{ba})t} - e^{-\Gamma t/2}}{\omega_\mu - \omega_{ba} + \mathrm{i}\Gamma/2}, \tag{15.42}$$

with $\omega_{ba} = \omega_b - \omega_a$ being the atomic level distance (in units of $\hbar$) or, in other words, the atomic resonance frequency. In the considered approximation, the interaction Hamiltonian $\hat{H}^{\mathrm{I}}$ reads

$$\hat{H}^{\mathrm{I}} = -\left[\hat{\mathbf{D}}^{(+)} \hat{\mathbf{E}}_{\mathrm{tot}}^{(-)}(0) + \hat{\mathbf{D}}^{(-)} \hat{\mathbf{E}}_{\mathrm{tot}}^{(+)}(0) \right]. \tag{15.43}$$

The operator $\hat{\mathbf{D}}$ is the atomic dipole operator, $\hat{\mathbf{E}}_{\mathrm{tot}}$ is the operator of the total electric field strength, and the $\pm$ sign indicates the positive and negative frequency parts. The electric field strength has to be taken at the origin of the coordinate system $\mathbf{r} = 0$ where the atom is located (the operators are time independent as we are working in the Schrödinger picture).

According to Equation (15.2), we have

$$\hat{\mathbf{E}}^{(+)}(0) = \sum_\mu \mathbf{e}_\mu \hat{a}_\mu, \quad \hat{\mathbf{E}}^{(-)}(0) = \sum_\mu \mathbf{e}_\mu^* \hat{a}_\mu^\dagger, \tag{15.44}$$

where for simplicity we have written $\mathbf{e}_\mu$ instead of $\mathbf{E}_\mu(0)$. Introducing the abbreviated notation $\mathbf{D}_{ab}$ for the matrix element $\langle a|\mathbf{D}^{(+)}|b\rangle$ and using Equation (15.5) the matrix element of the interaction Hamiltonian reads

$$\langle a, 0, 0, \ldots, 1_\mu, 0, \ldots |\hat{H}^I|b, 0, 0, \ldots\rangle = -\mathbf{D}_{ab}\mathbf{e}_\mu^*. \qquad (15.45)$$

The calculation shows that the (positive) constant Γ in Equations (15.41) and (15.42) is determined by the coupling parameters of Equation (15.45) in the form

$$\Gamma = 2i\hbar^{-2}\sum_\mu \frac{|\mathbf{D}_{ab}\mathbf{e}_\mu^*|^2}{\omega_{ba}-\omega_\mu+i\eta} \quad (\eta \to +0)$$

$$= 2\pi\hbar^{-2}\sum_p \int d\Omega |\mathbf{D}_{ab}\mathbf{e}_\mu^*|^2\rho(\omega)|_{\omega=\omega_{ba}}. \qquad (15.46)$$

Here, the density of radiation field states is denoted by $\rho(\omega)$, p is the polarization direction of the emitted plane wave and the solid angle Ω characterizes the propagation direction.

The Weisskopf–Wigner solution has the great advantage that it is not a perturbation theoretical approximation and hence is valid also for long times. It fulfils all expectations (see Section 6.5): it exhibits an exponential decay of the upper level population (Equation (15.41)); the emitted radiation has a Lorentz-like line shape (Equation (15.42)); and it satisfies the relation between the emission duration and the linewidth that is known from classical optics. Finally, for the damping constant Γ we obtain using Equation (15.46) the connection with the transition dipole matrix elements $\mathbf{D}_{ab}$ known from perturbation theory, which shows that the decay is faster the bigger the atomic dipole moment, i.e. the stronger the coupling.

15.4 Theory of beamsplitting and optical mixing

Let us assume that two optical waves 1 and 2 impinge on a lossless beamsplitter – a partially transmitting mirror (see Fig. 7.8). The energy conservation law requires that the incident energy is completely transferred into the output beams 3 and 4. When the two incoming waves, assumed to be plane waves for simplicity, have the same frequency, the classical description requires the intensities $A_i^*A_i$ (A_i being the amplitude of the respective wave) to fulfil the relation

$$A_1^*A_1 + A_2^*A_2 = A_3^*A_3 + A_4^*A_4. \qquad (15.47)$$

This conservation law is generally valid when the amplitudes of the incident waves are connected with those of the outgoing waves through a unitary transformation. Choosing the latter in the form

$$\begin{pmatrix} A_3 \\ A_4 \end{pmatrix} = \begin{pmatrix} \sqrt{t} & \sqrt{r} \\ -\sqrt{r} & \sqrt{t} \end{pmatrix}\begin{pmatrix} A_1 \\ A_2 \end{pmatrix}, \qquad (15.48)$$

we describe a normal type of mirror with reflectivity r and transmittivity t $(= 1 - r)$. The transmitted beam passes without a phase shift; the beam reflected on the one side of the mirror, however, acquires a phase jump of π.

The quantum mechanical description of the beamsplitter is obtained simply by replacing the classical (complex) amplitudes in Equation (15.48) by the corresponding photon annihilation operators

$$\begin{pmatrix} \hat{a}_3 \\ \hat{a}_4 \end{pmatrix} = \begin{pmatrix} \sqrt{t} & \sqrt{r} \\ -\sqrt{r} & \sqrt{t} \end{pmatrix} \begin{pmatrix} \hat{a}_1 \\ \hat{a}_2 \end{pmatrix}, \tag{15.49}$$

Apart from energy conservation, the unitarity matrix has another important function here: it guarantees the validity of the commutation relations

$$[\hat{a}_3, \hat{a}_4] = [\hat{a}_3^\dagger, \hat{a}_4^\dagger] = [\hat{a}_3, \hat{a}_4^\dagger] = [\hat{a}_4, \hat{a}_3^\dagger] = 0, \tag{15.50}$$

$$[\hat{a}_3, \hat{a}_3^\dagger] = [\hat{a}_4, \hat{a}_4^\dagger] = \mathbf{1}, \tag{15.51}$$

(see Equation (15.3)) for the outgoing waves, when the corresponding relations are satisfied by the incident waves. The unitarity ensures the consistency of the quantum mechanical description.

With the help of Equation (15.49), the process of beamsplitting can be described rather easily. Because the transformation is chosen to be real, it holds also for the photon creation operators. Inverting the relations we find

$$\begin{pmatrix} \hat{a}_1^\dagger \\ \hat{a}_2^\dagger \end{pmatrix} = \begin{pmatrix} \sqrt{t} & -\sqrt{r} \\ \sqrt{r} & \sqrt{t} \end{pmatrix} \begin{pmatrix} \hat{a}_3^\dagger \\ \hat{a}_4^\dagger \end{pmatrix}. \tag{15.52}$$

Let us assume that mode 1 is prepared in a state with exactly n photons and that mode 2 is empty. The initial state can be written, using the binomial theorem and Equations (15.6) and (15.52), in the form

$$|n\rangle_1 |0\rangle_2 |0\rangle_3 |0\rangle_4 = \frac{\hat{a}_1^{\dagger n}}{\sqrt{n!}} |0\rangle_1 |0\rangle_2 |0\rangle_3 |0\rangle_4$$

$$= \frac{1}{\sqrt{n!}} \sum_{k=0}^{n} \binom{n}{k} \sqrt{t}^k (-\sqrt{r})^{n-k} \hat{a}_3^{\dagger k} \hat{a}_4^{\dagger(n-k)} |0\rangle_1 |0\rangle_2 |0\rangle_3 |0\rangle_4$$

$$= \sum_{k=0}^{n} \sqrt{\binom{n}{k}} \sqrt{t}^k (-\sqrt{r})^{n-k} |0\rangle_1 |0\rangle_2 |k\rangle_3 |n-k\rangle_4. \tag{15.53}$$

The last expression gives us the wave function of the outgoing light. It is obviously an entangled state. Let us analyze the simplest case, namely that of a single photon

($n = 1$) impinging on a half transparent mirror; we learn from Equation (15.53) that the outgoing light is in the superposition state

$$|\psi^{(1)}\rangle = \frac{1}{\sqrt{2}}(|1\rangle_3|0\rangle_4 - |0\rangle_3|1\rangle_4), \tag{15.54}$$

which, after reuniting the two beams in an interferometer, gives rise to an interference effect that can be understood as a consequence of the indistinguishability of the paths the photon might have taken in the interferometer (see Section 7.2).

Performing measurements on only one of the beams, for example the transmitted beam 3, we obtain the corresponding density matrix by tracing out the unobserved sub-system in Equation (15.53):

$$\hat{\rho}_3 = \sum_{k=0}^{n} w_k^{(n)}|k\rangle_{33}\langle k| \tag{15.55}$$

with

$$w_k^{(n)} = \binom{n}{k} t^k r^{n-k}. \tag{15.56}$$

Obviously, Equation (15.55) represents a mixture, while the total system is in a pure state.

Due to the linearity of the beamsplitting process with the result in Equation (15.53), it is easy to predict how the beamsplitter transforms a general initial state. It induces the transformation

$$\sum_{n=0}^{\infty} c_n|n\rangle_1|0\rangle_2 \rightarrow \sum_{n=0}^{\infty} c_n|\Phi_n\rangle, \tag{15.57}$$

where we abbreviated the last line of Equation (15.53) by $|\Phi_n\rangle$ (the common product vector $|0\rangle_1|0\rangle_2$ was omitted).

For the physically important case that the incident beam is in a Glauber state $|\alpha\rangle$, the transformation simplifies to

$$|\alpha\rangle_1|0\rangle_2 \rightarrow |\sqrt{t}\alpha\rangle_3| - \sqrt{r}\alpha\rangle_4. \tag{15.58}$$

The outgoing light beams are – in complete agreement with the classical description – also in Glauber states (with correspondingly attenuated amplitudes). It is remarkable that the state of the outgoing light is not entangled. The result in Equation (15.58) has great practical importance. Utilizing attenuation, a conventional absorber might also be used instead of a beamsplitter; we can prepare arbitrarily weak Glauber light from intense Glauber light (laser radiation).

We should point out that we can easily derive a useful relation for the change of the photon number factorial moments,

$$M^{(j)} \equiv \overline{n(n-1)\cdots(n-j+1)} \quad (j = 1, 2, \ldots), \tag{15.59}$$

in transmission or reflection using the quantum mechanical formalism. These quantities are quantum mechanically simply the expectation values of the (normally ordered) operator products $\hat{a}^{\dagger j}\hat{a}^{j}$. Calculating them, for example for the transmitted wave, with the help of Equation (15.49) and the Hermitian conjugate equation under the assumption that only the first mode is excited, all the terms to which the second mode contributes vanish. What remains is the simple relation

$$M_3^{(j)} \equiv \langle \hat{a}_3^{\dagger j}\hat{a}_3^{j}\rangle = t^j \langle \hat{a}_1^{\dagger j}\hat{a}_1^{j}\rangle \equiv t^j M_1^{(j)}, \tag{15.60}$$

and an analogous equation holds true for the reflected wave (we have to replace the transmittivity t by the reflectivity r).

As we have seen, in the beamsplitter transformation of Equation (15.49) we must also take into account, for reasons of consistency, the second mode, even when it is "empty." By this we mean that the vacuum mode is coupled through the unused input port. The energy balance is not influenced, but formally the vacuum field gives rise to additional fluctuations of the radiation field; vacuum fluctuations are, so to speak, entering into the apparatus. This suggestive picture is especially useful when we do not count photons but measure the outcoming field with the help of the homodyne technique (see Section 10.2).

The beamsplitter can also be used as an *optical mixer*. To this end, light has to be sent also into the usually unused input port. The formal mathematical apparatus used to describe this mixing process is at hand in the form of Equation (15.49). The experimentally simply realizable case of exactly one photon entering each of the input ports is of particular interest (see Section 7.6). With the help of Equation (15.52) we find

$$\begin{aligned}
|1\rangle_1|1\rangle_2|0\rangle_3|0\rangle_4 &= \hat{a}_1^{\dagger}\hat{a}_2^{\dagger}|0\rangle_1|0\rangle_2|0\rangle_3|0\rangle_4 \\
&= \left\{ \sqrt{rt}\left(\hat{a}_3^{\dagger 2} - \hat{a}_4^{\dagger 2}\right) + (t-r)\hat{a}_3^{\dagger}\hat{a}_4^{\dagger} \right\} |0\rangle_1|0\rangle_2|0\rangle_3|0\rangle_4.
\end{aligned} \tag{15.61}$$

Specializing to the case of a balanced mirror ($t = r = 1/2$), we arrive at the following surprising result:

$$|1\rangle_1|1\rangle_2|0\rangle_3|0\rangle_4 = \frac{1}{\sqrt{2}}[|2\rangle_3|0\rangle_4 - |0\rangle_3|2\rangle_4]|0\rangle_1|0\rangle_2, \tag{15.62}$$

which is obviously telling us that the photons are "inseparable" once mixed: when we "look" at them we find them always both in one of the output ports but never one of them in each output. This means that the two detectors never indicate coincidences.

15.5 Quantum theory of interference

The basic principle of interference is that two (or more) optical fields are superposed. A detector placed at a position $\mathbf{r}$ reacts naturally to the total electric field strength $\mathbf{E}_{\text{tot}}(\mathbf{r}, t) = \mathbf{E}^{(1)}(\mathbf{r}, t) + \mathbf{E}^{(2)}(\mathbf{r}, t)$ residing on its sensitive surface. The response probability (per second) of the detector for quasimonochromatic light, according to quantum mechanics, is

$$W = \beta \langle \hat{\mathbf{E}}_{\text{tot}}^{(-)}(\mathbf{r}) \hat{\mathbf{E}}_{\text{tot}}^{(+)}(\mathbf{r}) \rangle, \tag{15.63}$$

(see Equation (5.4)), where the constant β is proportional to the detection efficiency

Let us denote (for reasons that will become clear later) the two beams that are made to interfere by 3 and 4; then Equation (15.63) takes the form

$$W = \beta \left\{ |\mathcal{E}_3(\mathbf{r})|^2 \langle \hat{a}_3^\dagger \hat{a}_3 \rangle + |\mathcal{E}_4(\mathbf{r})|^2 \langle \hat{a}_4^\dagger \hat{a}_4 \rangle \right.$$
$$\left. + \mathcal{E}_3^*(\mathbf{r}) \mathcal{E}_4(\mathbf{r}) \langle \hat{a}_3^\dagger \hat{a}_4 \rangle + \mathcal{E}_3(\mathbf{r}) \mathcal{E}_4^*(\mathbf{r}) \langle \hat{a}_3 \hat{a}_4^\dagger \rangle \right\}. \tag{15.64}$$

Here we have used Equation (15.2) and we have assumed both waves to be linearly (and identically) polarized. To keep the analysis simple, let us assume the waves to be plane waves (which, due to the presence of mirrors, change their directions before they become reunited). This means that the absolute values of $\mathcal{E}_3(\mathbf{r})$ and $\mathcal{E}_4(\mathbf{r})$ are the same and independent of $\mathbf{r}$. The product $\mathcal{E}_3^*(\mathbf{r}) \mathcal{E}_4(\mathbf{r})$ contains an additional phase factor $\exp(\mathrm{i}\Delta\varphi)$. The *classical* phase $\Delta\varphi$ is (apart from possible phase jumps) determined by the path difference ΔL, namely it is given by $\Delta\varphi = 2\pi \Delta L/\lambda$, with λ being the wavelength. Thus, Equation (15.64) simplifies to

$$W = \text{const} \times \left\{ \langle \hat{a}_3^\dagger \hat{a}_3 \rangle + \langle \hat{a}_4^\dagger \hat{a}_4 \rangle + \mathrm{e}^{\mathrm{i}\Delta\varphi} \langle \hat{a}_3^\dagger \hat{a}_4 \rangle + \mathrm{e}^{-\mathrm{i}\Delta\varphi} \langle \hat{a}_3 \hat{a}_4^\dagger \rangle \right\}. \tag{15.65}$$

As is well known, the condition *sine qua non* for the appearance of interference is the existence of a phase relation between the partial waves. We recognize this in Equation (15.65) from the fact that the interference terms are determined by the correlation terms $\langle \hat{a}_3^\dagger \hat{a}_4 \rangle = \langle \hat{a}_3 \hat{a}_4^\dagger \rangle^*$. In conventional interference experiments, the required phase relation is produced by splitting the primary beam either by beam or wavefront division. The result is two coherent beams, i.e. beams that can be made to interfere. The beamsplitter discussed theoretically in Section (15.4) is an appropriate model for the description of the division process. The expectation values appearing in Equation (15.65) can be easily expressed through the mean photon number $\overline{N}_1 = \langle \hat{a}_1^\dagger \hat{a}_1 \rangle$ of wave 1 assumed to be incident alone, using Equation (15.49) and the Hermitian conjugate relation. From energy conservation follows

$$\langle \hat{a}_3^\dagger \hat{a}_3 \rangle + \langle \hat{a}_4^\dagger \hat{a}_4 \rangle = \langle \hat{a}_1^\dagger \hat{a}_1 \rangle + \langle \hat{a}_2^\dagger \hat{a}_2 \rangle = \overline{N}_1. \tag{15.66}$$

The mixed terms are readily calculated with the help of the relation

$$\hat{a}_3^{\dagger}\hat{a}_4 = \sqrt{rt}\left(\hat{a}_2^{\dagger}\hat{a}_2 - \hat{a}_1^{\dagger}\hat{a}_1\right) - r\hat{a}_2^{\dagger}\hat{a}_1 + t\hat{a}_1^{\dagger}\hat{a}_2, \tag{15.67}$$

and because mode 2 is in the vacuum state we obtain

$$\langle\hat{a}_3^{\dagger}\hat{a}_4\rangle = \langle\hat{a}_3\hat{a}_4^{\dagger}\rangle^* = -\sqrt{rt}\,\overline{N}_1. \tag{15.68}$$

Thus, Equation (15.65) takes the simple form

$$W = \text{const} \times \overline{N}_1\left(1 - 2\sqrt{rt}\cos\Delta\varphi\right). \tag{15.69}$$

This result coincides *exactly* with the classical formula. For the case of a balanced beamsplitter, the visibility of the interference pattern attains unity. The point is that we have reproduced this result quantum mechanically for an *arbitrary* input state. The quantum mechanical description of a conventional interference experiment does not reveal anything new. This holds true independently of the intensity of the incident wave, and in particular for arbitrarily weak intensities (see Equation (15.69), which helps us to understand Dirac's statement that we are always dealing with the "interference of a photon with itself." (We have to keep in mind that in the case $\overline{N}_1 \ll 1$ only those members of the ensemble described by the wave function or the density matrix contribute to the interference pattern on which the detector indeed registers a photon).

The situation changes when we consider the interference of independent light waves (coming from two independent lasers, for example). In this case, the expectation values in Equation (15.64) can be factorized as $\langle\hat{a}_3^{\dagger}\hat{a}_4\rangle = \langle\hat{a}_3^{\dagger}\rangle\langle\hat{a}_4\rangle$ and $\langle\hat{a}_3\hat{a}_4^{\dagger}\rangle = \langle\hat{a}_3\rangle\langle\hat{a}_4^{\dagger}\rangle$, respectively. Interference can now be observed only when the expectation values of $\hat{a}$ are non-zero for both waves, i.e. when the waves have a more or less well defined *absolute* phase. This requirement is definitely met by Glauber light. Choosing the radiation field to be in the state $|\psi\rangle = |\alpha_3\rangle_3|\alpha_4\rangle_4$, we can easily calculate the detector response probability with the help of Equations (15.24) and (15.25), the result being

$$W = \beta\{|\mathcal{E}_3(\mathbf{r})|^2|\alpha_3|^2 + |\mathcal{E}_4(\mathbf{r})|^2|\alpha_4|^2 + \mathcal{E}_3^*(\mathbf{r})\mathcal{E}_4(\mathbf{r})\alpha_3^*\alpha_4 + \mathcal{E}_3(\mathbf{r})\mathcal{E}_4^*(\mathbf{r})\alpha_3\alpha_4^*\}. \tag{15.70}$$

Specializing as before to the case of a linearly polarized plane wave, we can simplify the previous equation to

$$W = \text{const} \times \left[|\alpha_3|^2 + |\alpha_4|^2 + e^{i\Delta\varphi}\alpha_3^*\alpha_4 + e^{-i\Delta\varphi}\alpha_3\alpha_4^*\right], \tag{15.71}$$

and we arrive at the same expression as in classical theory. This is not surprising when we recall the general correspondence between the classical and the quantum mechanical description stated in Section 15.2. However, it is amazing that the visibility of the interference pattern does not change even for *arbitrarily* weak intensities and equals unity whenever the (mean!) photon numbers $|\alpha_3|^2$ and $|\alpha_4|^2$ are equal.

15.6 Theory of balanced homodyne detection

The homodyne technique described in Section 9.3 (see also Fig. 9.1) can be easily treated theoretically when we also describe the photocurrent quantum mechanically. In the detection process (we consider 100% efficiency detectors), each photon is converted into an electron, and so it appears natural to identify the photocurrent, apart from a factor e (elementary charge), with the photocurrent also in the sense of an operator relation. In the case of a quasimonochromatic wave, we thus find the simple expression for the photocurrent to be

$$\hat{I} = s\hat{a}^\dagger\hat{a}, \tag{15.72}$$

where $\hat{a}^\dagger$ and $\hat{a}$ are the creation and annihilation operators (see Section 15.1) and s is a constant (ensuring in particular the dimensional correctness of the relation).

Using a balanced beamsplitter (see Fig. 9.1), signal 1 is mixed with a local oscillator. This process is described by Equation (15.49). Assuming the local oscillator to be in a Glauber state $|\alpha_L\rangle$ (in practice a laser beam) with a large amplitude, we can approximate the operators $\hat{a}_L$ and $\hat{a}_L^\dagger$ by complex numbers α_L and α_L^*. The photocurrent operators $\hat{I}_3$ and $\hat{I}_4$ take the form

$$\hat{I}_3 = \tfrac{1}{2}s(\hat{a}_1^\dagger + \alpha_L^*)(\hat{a}_1 + \alpha_L), \tag{15.73}$$

$$\hat{I}_4 = \tfrac{1}{2}s(-\hat{a}_1^\dagger + \alpha_L^*)(-\hat{a}_1 + \alpha_L). \tag{15.74}$$

Subtracting these two equations, we obtain the simple relation

$$\Delta\hat{I} \equiv \hat{I}_3 - \hat{I}_4 = s(\hat{a}_1^\dagger\alpha_L + \hat{a}_1\alpha_L^*) \tag{15.75}$$

for the actually measured quantity.

Introducing the phase of the local oscillator through the relation $\alpha_L = |\alpha_L|\exp(i\Theta)$, we realize that we measure the observable

$$\hat{x}_\Theta \equiv \frac{1}{\sqrt{2}}\left(e^{i\Theta}\hat{a}_1^\dagger + e^{-i\Theta}\hat{a}_1\right), \tag{15.76}$$

and this is, according to Equation (15.12), simply a special quadrature component. Through the variation of Θ for an unchanged signal, it is possible to measure a whole set of quadrature components $\hat{x}_\Theta$. This is what we need for the tomographic reconstruction of the Wigner function (see Section 10.3).

References

Alguard, M. J. and C. W. Drake. 1973. *Phys. Rev. A* **8**, 27.

Arecchi, F. T., A. Berne and P. Burlamacchi. 1966. *Phys. Rev. Lett.* **16**, 32.

Arecchi, F. T., V. Degiorgio and B. Querzola. 1967a. *Phys. Rev. Lett.* **19**, 1168.

Arecchi, F. T., M. Giglio and U. Tartari. 1967b. *Phys. Rev.* **163**, 186.

Arnold, W., S. Hunklinger and K. Dransfeld. 1979. *Phys. Rev. B* **19**, 6049.

Aspect, A., P. Grangier and G. Roger. 1981. *Phys. Rev. Lett.* **47**, 460.

Aspect, A., J. Dalibard and G. Roger. 1982. *Phys. Rev. Lett.* **49**, 1804.

Bandilla, A. and H. Paul. 1969. *Ann. Physik* **23**, 323.

Basché, T., W. E. Moerner, M. Orrit and H. Talon. 1992. *Phys. Rev. Lett.* **69**, 1516.

Bell, J. S. 1964. *Physics* **1**, 195.

Bennett, C. H., G. Brassard and A. K. Ekert. 1992. *Sci. Am.* **26**, 33.

Bennett, C. H., G. Brassard, C. Crépeau, R. Josza, A. Peres and W. K. Wootters. 1993. *Phys. Rev. Lett.* **70**, 1895.

Beth, R. A. 1936. *Phys. Rev.* **50**, 115.

Born, M. and E. Wolf. 1964. *Principles of Optics*, 2nd edn. Oxford: Pergamon Press, p. 10.

Bouwmeester, D., J.-W. Pan, K. Mattle, M. Eibl, H. Weinfurter and A. Zeilinger. 1997. *Nature* **390**, 575.

Braunstein, S. L. and A. Mann. 1995. *Phys. Rev. A* **51**, R1727.

Braunstein, S. L. and H. J. Kimble. 1998. *Phys. Rev. Lett.* **80**, 869.

Brendel, J., S. Schütrumpf, R. Lange, W. Martienssen and M. O. Scully. 1988. *Europhys. Lett.* **5**, 223.

Brunner, W., H. Paul and G. Richter. 1964. *Ann. Physik* **14**, 384. 1965. *Ann. Physik* **15**, 17.

Clauser, J. F. and A. Shimony. 1978. *Rep. Prog. Phys.* **41**, 1881.

Clauser, J. F., M. A. Horne, A. Shimony and R. A. Holt. 1969. *Phys. Rev. Lett.* **23**, 880.

Dagenais, M. and L. Mandel. 1978. *Phys. Rev. A* **18**, 2217.

Davis, C. C. 1979. *IEEE J. Quantum Electron.* **QE-15**, 26.

De Martini, F., G. Innocenti, G. R. Jacobovitz and P. Mataloni. 1987. *Phys. Rev. Lett.* **26**, 2955.

Dempster, A. J. and H. F. Batho. 1927. *Phys. Rev.* **30**, 644.

Diedrich, F. and H. Walther. 1987. *Phys. Rev. Lett.* **58**, 203.

Dirac, P. A. M. 1927. *Proc. Roy. Soc. (London) A* **114**, 243. 1958. *The Principles of Quantum Mechanics*, 4th edn. London: Oxford University Press, p. 9.

Drexhage, K. H. 1974. Interaction of light with monomolecular dye layers. In E. Wolf (ed.), *Progress in Optics*, vol. 12. Amsterdam: North-Holland Publishing Company, p. 163.

Dürr, S., T. Nonn and G. Rempe. 1998. *Nature* **395**, 33.

Einstein, A. 1905. *Ann. Physik* **17**, 132.

    1906. *Ann. Physik* **20**, 199.

    1917. *Phys. Ztschr.* **18**, 121.

Einstein, A., B. Podolsky and N. Rosen. 1935. *Phys. Rev.* **47**, 777.

Foord, R., E. Jakeman, R. Jones, C. J. Oliver and E. R. Pike. 1969. *IERE Conference Proceedings No. 14.*

Forrester, A. T., R. A. Gudmundsen and P. O. Johnson. 1955. *Phys. Rev.* **99**, 1691.

Franck, J. and G. Hertz. 1913. *Verh. d. Dt. Phys. Ges.* **15**, 34, 373, 613, 929.

    1914. *Verh. d. Dt. Phys. Ges.* **16**, 12, 457, 512.

Franson, J. D. 1989. *Phys. Rev. Lett.* **62**, 2205.

Freedman, S. J. and J. F. Clauser. 1972. *Phys. Rev. Lett.* **28**, 938.

Freyberger, M., K. Vogel and W. Schleich. 1993. *Quantum Opt.* **5**, 65.

Furusawa, A., J. L. Sørensen, S. L. Braunstein, C. A. Fuchs, H. J. Kimble and E. S. Polzik. 1998. *Science* **282**, 706.

Ghosh, R. and L. Mandel. 1987. *Phys. Rev. Lett.* **59**, 1903.

Glauber, R. J. 1965. Optical coherence and photon statistics. In C. De Witt, A. Blandin and C. Cohen-Tannoudji (eds.), *Quantum Optics and Electronics*. New York: Gordon and Breach.

    1986. Amplifiers, attenuators and the quantum theory of measurement. In E. R. Pike and S. Sarkar (eds.), *Frontiers in Quantum Optics*. Bristol: Adam Hilger Ltd.

Gordon, J. P., H. J. Zeiger and C. H. Townes. 1954. *Phys. Rev.* **95**, 282.

Grishaev, I. A., I. S. Guk, A. S. Mazmanishvili and A. S. Tarasenko. 1972. *Zurn. eksper. teor. Fiz.* **63**, 1645.

Hanbury Brown, R. 1964. *Sky and Telescope* **28**, 64.

Hanbury Brown, R. and R. Q. Twiss. 1956a. *Nature* **177**, 27.

    1956b. *Nature* **178**, 1046.

Hauser, U., N. Neuwirth and N. Thesen. 1974. *Phys. Lett.* **49 A**, 57.

Heitler, W. 1954. *The Quantum Theory of Radiation*, 3rd edn. London: Oxford University Press.

Hellmuth, T., H. Walther, A. Zajonc and W. Schleich. 1987. *Phys. Rev. A* **35**, 2532.

Hong, C. K. and L. Mandel. 1986. *Phys. Rev. Lett.* **56**, 58.

Hong, C. K., Z. Y. Ou and L. Mandel. 1987. *Phys. Rev. Lett.* **59**, 2044.

Hulet, R. G., E. S. Hilfer and D. Kleppner. 1985. *Phys. Rev. Lett.* **55**, 2137.

Huygens, C. 1690. *Traité de la Lumière*. Leiden: Pierre von der Aa.

Itano, W. M., J. C. Bergquist, R. G. Hulet and D. J. Wineland. 1987. *Phys. Rev. Lett.* **59**, 2732.

Itano, W. M., D. J. Heinzen, J. J. Bollinger and D. J. Wineland. 1990. *Phys. Rev. A* **41**, 2295.

Jakeman, E., E. R. Pike, P. N. Pusey and J. M. Vaughan. 1977. *J. Phys. A* **10**, L 257.

Jánossy, L. 1973. Experiments and theoretical considerations concerning the dual nature of light. In H. Haken and M. Wagner (eds.), *Cooperative Phenomena*. Berlin: Springer-Verlag, p. 308.

Jánossy, L. and Z. Náray. 1957. *Acta Phys. Acad. Sci. Hung.* **7**, 403.

    1958. *Nuovo Cimento*, Suppl. **9**, 588.

Javan, A., E. A. Ballik and W. L. Bond. 1962. *J. Opt. Soc. Am.* **52**, 96.

Kimble, H. J., M. Dagenais and L. Mandel. 1977. *Phys. Rev. Lett.* **39**, 691.

Kwiat, P. G., A. M. Steinberg and R. Y. Chiao. 1993. *Phys. Rev. A* **47**, R 2472.
  1994. *Phys. Rev. A* **49**, 61.
Kwiat, P. G., K. Mattle, H. Weinfurter, A. Zeilinger, A. V. Sergienko and Y. H. Shih.
  1995. *Phys. Rev. Lett.* **75**, 4337.
Landau, L. D. and E. M. Lifschitz. 1965. *Lehrbuch der theoretischen Physik*, vol. III,
  Quantenmechanik. Berlin: Akademie-Verlag, p. 158.
Lawrence, E. O. and J. W. Beams. 1927. *Phys. Rev.* **29**, 903.
Lebedew, P. N. 1910. *Fis. Obosrenie* **11**, 98.
Lenard, P. 1902. *Ann. Physik.* **8**, 149.
Lenz, W. 1924. *Z. Phys.* **25**, 299.
Leonhardt, U. 1997. *Measuring the Quantum State of Light*. Cambridge: Cambridge
  University Press.
Leonhardt, U. and H. Paul. 1993a. *Phys. Rev. A* **47**, R 2460.
  1993b. *Phys. Rev. A* **48**, 4598.
  1995. *Prog. Quantum Electron.* **19**, 89.
London, F. 1926. *Z. Phys.* **40**, 193.
Lorentz, H. A. 1910. *Phys. Ztschr.* **11**, 1234.
Machida, S., Y. Yamamoto and Y. Itaya. 1987. *Phys. Rev. Lett.* **58**, 1000.
Magyar, G. and L. Mandel. 1963. *Nature* **198** (4877), 255.
Mandel, L. 1963. Fluctuations of light beams. In E. Wolf (ed.), *Progress in Optics*, vol.2.
  Amsterdam: North-Holland Publishing Company, p. 181.
  1976. *J. Opt. Soc. Am.* **66**, 968.
  1983. *Phys. Rev. A* **28**, 929.
Mandel, L. and E. Wolf. 1965. *Rev. Mod. Phys.* **37**, 231.
Martienssen, W. and E. Spiller. 1964. *Am. J. Phys.* **32**, 919.
Meschede, D., H. Walther and G. Müller. 1984. *Phys. Rev. Lett.* **54**, 551.
Michelson, A. A. and F. G. Pease. 1921. *Astrophys. J.* **53**, 249.
Middleton, D. 1960. *An Introduction to Statistical Communication Theory*. New York:
  McGraw-Hill.
Millikan, R. A. 1916. *Phys. Rev.* **7**, 373.
Neuhauser, W., M. Hohenstatt, P. E. Toschek and H. Dehmelt. 1980. *Phys. Rev. A* **22**,
  1137.
Newton, I. 1730. *Opticks: Or a Treatise of the Reflections, Refractions, Inflections
  and Colours of Light*, 4th edn. London. Reprinted 1952, New York: Dover
  Publications.
Noh, J. W., A. Fougéres and L. Mandel. 1991. *Phys. Rev. Lett.* **67**, 1426.
  1992. *Phys. Rev. A* **45**, 424.
Osnaghi, S., P. Bertet, A. Auffeves, P. Maioli, M. Brune, J. M. Raimond and S. Haroche.
  2001. *Phys. Rev. Lett.* **87**, 037902–1.
Ou, Z. Y., L. J. Wang and L. Mandel. 1989. *Phys. Rev. A* **40**, 1428.
Paul, H. 1966. *Fortschr. Phys.* **14**, 141.
  1969. *Lasertheorie I*. Berlin: Akademie-Verlag.
  1973a. *Nichtlineare Optik I*. Berlin: Akademie-Verlag.
  1973b. *Nichtlineare Optik II*. Berlin: Akademie-Verlag.
  1974. *Fortschr. Phys.* **22**, 657.
  1980. *Fortschr. Phys.* **28**, 633.
  1981. *Opt. Acta* **28**, 1.
  1985. *Am. J. Phys.* **53**, 318.
  1986. *Rev. Mod. Phys.* **58**, 209.
  1996. *Opt. Quantum Electron.* **28**, 1111.

Paul, H., W. Brunner and G. Richter. 1963. *Ann. Physik.* **12**, 325.

    1965. *Ann. Physik.* **16**, 93.

Paul, H. and R. Fischer. 1983. *Usp. fiz. nauk.* **141**, 375.

Pauli, W. 1933. Die allgemeinen Prinzipien der Wellenmechanik. In H. Geiger and
    K. Scheel (eds.), *Handbuch der Physik* 24/1, 2nd edn. Berlin: Springer-Verlag.
    English translation: Pauli, W. 1980. *General Principles of Quantum Mechanics*.
    Berlin: Springer-Verlag, p. 21.

Pease, F. G. 1931. *Ergeb. exakt. Naturwissensch.* **10**, 84.

Pfleegor, R. L. and L. Mandel. 1967. *Phys. Rev.* **159**, 1084.

    1968. *J. Opt. Soc. Am.* **58**, 946.

Pike, E. R. 1970. Photon statistics. In S. M. Kay and A. Maitland (eds.), *Quantum Optics*.
    London: Academic Press.

Planck, M. 1943. *Naturwissensch.* **31**, 153.

    1966. *Theorie der Wärmestrahlung*, 6th edn. Leipzig: J. A. Barth, pp. 190 ff.

Power, E. A. 1964. *Introductory Quantum Electrodynamics*. London: Longmans.

Prodan, J.V., W. D. Phillips and H. Metcalf. 1982. *Phys. Rev. Lett.* **49**, 1149.

Radloff, W. 1968. *Phys. Lett. A* **27**, 366.

    1971. *Ann. Physik.* **26**, 178.

Rebka, G. A. and R. V. Pound. 1957. *Nature* **180**, 1035.

Rempe, G., F. Schmidt-Kaler and H. Walther. 1990. *Phys. Rev. Lett.* **64**, 2783.

Renninger, M. 1960. *Ztschr. Phys.* **158**, 417.

Reynolds, G. T., K. Spartalian and D. B. Scarl. 1969. *Nuovo Cimento* **61 B**, 355.

Roditschew, W. I. and U. I. Frankfurt (eds.) 1977. *Die Schöpfer der physikalischen Optik*.
    Berlin: Akademie-Verlag.

Sauter, T., R. Blatt, W. Neuhauser and P. E. Toschek. 1986. *Opt. Commun.* **60**, 287.

Schrödinger, E. 1935. *Proc. Camb. Phil. Soc.* **31**, 555.

Shapiro, J. H. and S. S. Wagner. 1984. *IEEE J. Quantum Electron.* **QE-20**, 803.

Sigel, M., C. S. Adams and J. Mlynek. 1992. Atom optics. In T. W. Haensch and
    M. Inguscio (eds.), *Frontiers in Laser Spectroscopy*. Proceedings of the International
    School of Physics 'Enrico Fermi,' Course CXX, Varenna.

Sillitoe, R. M. 1972. *Proc. Roy. Soc. Edinburgh, Sect. A*, **70 A**, 267.

Sleator, T., O. Carnal, T. Pfau, A. Faulstich, H. Takuma and J. Mlynek. 1992. In
    M. Ducloy *et al.* (eds.), *Proceedings of the Tenth International Conference on Laser
    Spectroscopy*. Singapore: World Scientific, p. 264.

Slusher, R. E., L. W. Hollberg, B. Yurke, J. C. Mertz and J. F. Valley. 1985. *Phys. Rev.
    Lett.* **55**, 2409.

Slusher, R. E. and B. Yurke. 1986. Squeezed state generation experiments in an optical
    cavity. In E. R. Pike and S. Sarkar (eds.), *Frontiers in Quantum Optics*, Bristol:
    Adam Hilger Ltd.

Smithey, D. T., M. Beck, M. G. Raymer and A. Faridani. 1993. *Phys. Rev. Lett.* **70**, 1244.

Sommerfeld, A. 1949. *Vorlesungen über Theoretische Physik*, vol. III, *Elektrodynamik*.
    Leipzig: Akademische Verlagsgesellschaft.

    1950. *Vorlesungen über Theoretische Physik*, vol. IV, *Optik*. Wiesbaden: Dieterich'sche
    Verlagsbuchhandlung.

Taylor, G. I. 1909. *Proc. Camb. Phil. Soc.* **15**, 114.

Teich, M. C. and B. E. A. Saleh. 1985. *J. Opt. Soc. Am. B* **2**, 275.

Tittel, W., J. Brendel, H. Zbinden and N. Gisin. 1998. *Phys. Rev. Lett.* **81**, 3563.

Vaidman, L. 1994. *Phys. Rev. A* **49**, 1473.

Vogel, K. and H. Risken. 1989. *Phys. Rev. A* **40**, 2847.

Walker, J. G. and E. Jakeman. 1985. *Opt. Acta* **32**, 1303.

Wawilow, S. I. 1954. *Die Mikrostruktur des Lichtes*. Berlin: Akademie-Verlag.

Weisskopf, V. and E. Wigner. 1930a. *Z. f. Phys.* **63**, 54.
    1930b. *Z. f. Phys.* **65**, 18.

Weihs, G., T. Jennewein, C. Simon, H. Weinfurter and A. Zeilinger. 1998. *Phys. Rev. Lett.* **81**, 5039.

Wootters, W. K. and W. H. Zurek. 1982. *Nature* **299**, 802.

Wu, L.-A., H. J. Kimble, J. L. Hall and H. Wu. 1986. *Phys. Rev. Lett.* **57**, 2520.

Young, T. 1802. *Phil. Trans. Roy. Soc. London* **91**, part 1, 12.
    1807. *Lectures on Natural Philosophy*, vol. 1. London.

Zou, X. Y., L. J. Wang and L. Mandel. 1991. *Phys. Rev. Lett.* **67**, 318.

Index

Printed in the United States
By Bookmasters